Agriculture and Rural Poverty

Edited & Compiled By

S.P. Malik

Educational Publishers & Distributors
291, Bank Enclave, Laxmi Nagar, Delhi-110092.

Published by:

Educational Publishers & Distributors

291, Bank Enclave, Laxmi Nagar, Delhi-110092.

Ph.: (O) 011-65190343, 43551324, 42564726
(M) 9811088729, (Fax) 42564726, (Res.) 47091605

E-mail: *khelsahitya1@rediffmail.com*

Web: *www.khelsahitya.com*

I.S.B.N. - 978-93-80873-37-4

PRINTED IN INDIA 2013

Laser Typeset by:

JAIN MEDIA GRAPHICS
Delhi-110035.

Printed by:

ASIAN OFFSET PRINTERS
Delhi

Price: Rs. 650/-

agricultural performance. A summary view of the evidence on changes in rural poverty and factors affecting these changes is presented in Section V.

The Poverty Line in Current Prices

The first step in our analysis is the definition of an appropriate poverty line for measuring absolute poverty. A fundamental limitation of this approach is that any such line is necessarily arbitrary. In this paper we finesse the problem by choosing our poverty line primarily to conform to past practice, without attempting to justify it as measuring some objectively defensible minimum standard. Accordingly, the poverty line used throughout this paper is a consumer expenditure level of Rs. 15 per person for 30 days at 1960-1 rural prices This line has a well-established pedigree in the Indian literature. As shown by Bardhan an expenditure level of Rs. 15 in 1960-1 at rural prices roughly corresponds to Rs. 20 per person at all-India 1960-1 prices, which is the minimum level originally adopted by the Planning Commission in 1962. Dandekar and Rath also adopted this line on the grounds that it corresponded to the expenditure level at which food consumption provided the 'norm' of 2250 calories per day. However, it is important to emphasise that attempts to interpret this line as guaranteeing a nutritional minimum could be seriously misleading. Suffice to say that this level of expenditure represents an extremely low level of living and one that has been widely accepted as a 'minimum level' in the policy debate.

Since commodity prices vary significantly across states, the same real consumption level requires different levels of consumer expenditure across states Bardhan estimated the level of consumption expenditure for each state in 1960-1, which is equivalent to consumption expenditure of Rs. 15 at all-India rural prices. We have adopted these estimates as our base year estimates of the poverty line in each state.

The second step in our analysis is the definition of equivalent poverty lines for different years in terms of consumer expenditure in the current prices for each year. This requires identification of a suitable price index for the rural poor. Price indices for the 'average' consumer are clearly not suitable since the poor spend a much greater proportion of their budget on items whose prices displayed very high inflation rates in the 'sixties e.g. food and especially coarse grains. In the absence of price indices specially designed for the rural poor, we have followed Bardhan in using the Consumer Price Indices for Agricultural Labourers, prepared by the Labour Bureau, as the most appropriate for our purpose. These indices are available for India as a whole and separately for each state.[7]

Applying the CPIAL to the base year estimates of the poverty line in terms of consumption expenditure per person in 1960-1 prices, we can calculate equivalent poverty lines for each of the years for which NSS consumption distributions are available. The resulting estimates of the poverty line for each state, and for India as a whole, for fourteen years spanning the period from the late 'fifties to the early seventies. These estimates are obviously subject to all the limitations arising from the use of the CPIAL as the price index. Ideally, we should use separate price indices for different groups comprising the poor, especially distinguishing landless labourers relying on wage income from subsistence farmers who rely upon own consumption.[8] More seriously, it can be argued that the very approach of using a base weighted price index is flawed since it cannot reflect the impact of changing relative prices upon the commodity composition of consumption.[9] In defence of our estimates we can only state that they are certainly in line with past practice, and are probably the best estimates possible given available data.

Trends in Rural Poverty 1956-57 to 1973-74

The poverty lines presented have been used in conjunction with the NSS consumption distributions to estimate our two alternative poverty measures: the percentage of the rural population below the poverty line and the Sen Poverty Index These estimates are obtained from a two-stage procedure.

First, we estimate the Lorenz curve of the consumption distribution for each year using the method of Kakwani and Podder. The estimated parameters of the Lorenz curves are then used to obtain point estimates of the two poverty measures. This procedure has been used to estimate the incidence of poverty over time for rural India as a whole as well as for each state separately.

(a) All-India Results

Our estimates of the two poverty measures for rural India as a whole are presented in Table2. The NSS data permit two different estimates of the percentage of the rural population in poverty in India as a whole. Estimate I is obtained by applying the all-India poverty line for various years to the NSS consumption distribution for rural India. Estimate II is obtained as a weighted sum of the estimated percentages in poverty in individual states, obtained from the NSS distributions for individual states and the state specific poverty line. As there were substantial interstate differences in prices in the base year, and furthermore, inflation occurred at different rate across states, it can be argued that Estimate II, which based on state specific poverty lines, is a better estimate of the percentage of the rural population in poverty.

The most important feature of the results presented in Table 2 is the marked fluctuation over time in the extent or incidence of rural poverty. The percentage in poverty declines initially from over 50 per cent in the mid-fifties to around 40 per cent in 1960-61, rises sharply through the mid-sixties, reaching a peak in 1967-8, and then declines again. The Sen Index also displays the same pattern. Since this index reflects not only the percentage below the poverty line, but also the average shortfall of this group from the poverty line, it is reasonable to conclude from the range of variation in this index that we are measuring substantial fluctuations in the intensity of poverty and not merely marginal shifts of large numbers from a position slightly above the poverty line to a position slightly below.

It is important to determine whether the observed fluctuations arise solely from the sampling variation over time in our estimates, or whether they reflect genuine changes in the incidence of poverty arising from underlying economic factors. Fortunately, NSS surveys are conducted on the basis of interpenetrating sub-samples and difference between estimates of the incidence of poverty based on different sub-samples for the same year provide an indication of the range of variation in sample estimates. Separate tabulations of the data by sub-sample are available for the earlier years up to 1965-6 and have been used to obtain sub-sample estimates of each of our poverty measures for India as a whole. The range of variation between sub-sample estimates for the same year is clearly much smaller than the variations observed over time. This suggests that the measured fluctuation in the incidence of poverty reflects real changes in the severity of the poverty problem over time.

The existence of fluctuations over time implies that we cannot generalise about underlying trends on the basis of comparisons between selected endpoints. For example, Bardhan reported a sharp increase in the incidence of rural poverty between 1960-1 and 1968-9 while Lai has argued that the incidence of rural poverty declined between 1956-7 and 1970-1. Our estimates for these years derived by a different estimation method confirm the direction of change between these particular endpoints as reported in each study, but they also point to the danger of using such comparisons for any assessment of underlying trends. This can only be done on the basis of the time series as a whole.

These results provide no evidence for asserting a trend increase or decrease in rural poverty over the period as a whole The increasing incidence of poverty over the sixties reported by Bardhan appears as an upswing in a pattern of cyclical variation We note that our conclusion also differs from 1970-1. who reported a decline in the percentage of the rural population m poverty over the period to 1956-7 to 1967-8. This difference is due to identifiable differences on some key issues in estimating rural poverty. The absence of any discernible trend in the incidence poverty obviously implies an increase in the absolute numbers of people in poverty because of the growth in rural population over the period he absolute se of the poor population also fluctuates over the period.

In general, the time pattern of the incidence of poverty in individual states follows the pattern of fluctuation described for India as a whole. The incidence of poverty for almost all states, using either of the two poverty measures, declines up to the early sixties and then begins to rise again, reaching a peak, in 1967-8 or 1968-69, and declining again thereafter.

Once again, we have tested for the existence of an underlying trend by fitting a linear time trend to each of our poverty measures for individual states. The results are summarised. Only two states Assam and West Bengal show a significant trend increase in poverty, according to both our poverty measures Andhra Pradesh and Tamil Nadu show a trend decline in the incidence of poverty although in Tamil Nadu this is significant only when we use the Sen Index. For all other slates there is no significant time trend in the incidence of rural poverty using either measure.

Our results for Punjab and Haryana are particularly worth noting in view of the findings of other studies that the incidence of rural poverty has increased in this region, despite the visible success of the Green Revolution in raising output. For example. Bardhan found that the incidence of poverty in Punjab and Haryana increased between 1960-1 and 1967-8 and Rajaraman reported that the incidence of poverty in Punjab excluding Haryana increased between 1960-1 and 1970-1. We find that while both conclusions stand as long as we compare these particular endpoints, they are misleading in respect of underlying trends. The incidence of rural poverty in Punjab and Haryana appears to have been unusually low in 1960-1 and estimates for this year therefore should not be used as the base for comparison. Consideration of the time series as a whole clearly does not suggest a trend increase in the incidence

Ideally, the observed changes in the incidence of poverty over different sub-periods should be explained in terms of some explicit model of the determinants of rural poverty and the behaviour of these determinants over the sub-periods In this context, it is obviously relevant to consider the rate of agricultural growth, the factors determining its distribution across farm sizes, its impact on tenancy conditions, its effect on the demand for labour. etc. A complete exploration of the impact of these factors on the economic condition of different socioeconomic groups in the rural economy is obviously beyond the scope of this paper. Instead, we shall confine ourselves to examining the relationship between the incidence of poverty and some aggregative indices of agricultural performance over the period 1956-7to 1973-4. Once again the analysis is presented separately for India as a whole and for the individual states.

The All-India Evidence

We begin by postulating that an important determinant of the extent of rural poverty is the level of agricultural production relative to the size of the rural population Agriculture is not the only source of income in rural areas but it is the dominant source, and besides, the scale of non-agricultural income generating activity in rural areas almost certainly depends upon the level of agricultural production. If there is and trickle-down' mechanism at work in the rural economy we should expect increases in agricultural production per head to reduce the incidence of absolute poverty. Does the available evidence support this view?

The first point to note about the Indian experience of the past two decades is that the growth of agricultural output has only just kept pace with the growth of the rural population The result is a stagnation in output per head, which can be seen from the following results reporting linear lime trends titled to three different indices of output per head of the rural population.

The relationship between the incidence of poverty and agricultural performance over the period under review can be more systematically examined through regression analysis When this is done we find that improved agricultural performance is definitely associated with reductions in the incidence of poverty Table 5 presents the results of some regressions along these lines using three alternative dependent variables Estimate I and Estimate II of the percent of population in poverty and the Sen Index to measure poverty Initially we hypothesise that the incidence of poverty depends upon the level of NDP in agriculture per head of the rural population . As shown in Table 5, the coefficient on this variable is negative and significant in all cases. An alternative hypothesis is that the incidence of poverty depends not only on the current year's level of NDPARP but also on the level in the previous year. This is tested by using the average value of NDPARP for the current and previous year as the independent variable in the regressions We use the average value of the current and previous year rather than introduce both as independent variables solely because of the limited sample size.

It can be argued that while NDPARP is inversely related to the incidence of poverty, there are other factors operating in the rural economy leading to an increase in poverty incidence over time I his hypothesis can be crudely tested by including time as an additional explanatory variable as in Eqs. 1c, 2c, 3c. We find that the coefficient on this variable is not significant in am equation, suggesting that there is no underlying time trend in the incidence of poverty after allowing for changes in poverty incidence associated with changes in NDPARP.

What can we legitimately inter from these results? There is clear evidence of an inverse relationship between rural poverty and agricultural performance. As show n in Figure 1, fluctuation in poverty incidence simply mirrors the movement in agricultural production per head and this inverse relationship is even more firmly identifiable when account is taken of the lags involved. Such empirical relationships are at best a crude basis fur drawing inferences about complex causal mechanisms, but taken at face value they do suggest that there is some trickling down of benefits from increases in agricultural production.

The Evidence for Individual States

Does the inverse relationship between poverty and agricultural growth also hold at the level of individual states? Testing for such a relationship at the state level is subject to two important limitations. The first is conceptual: the rural area of a single state is a more open' economic system than rural India as a whole and therefore all-India relationships may not hold for individual states. For example, the effect of adverse agricultural conditions in a single state especially in a particular year on rural poverty in that state may be mitigated by temporary migration both to urban areas and to rural labour markets in other states. These 'cushioning' possibilities do not exist at the all-India level. The only way migration can mitigate the adverse effects of poor agricultural performance for the country as a whole is through migration to urban areas and that too is limited by the fact that a generally poor agricultural performance typically also has an adverse effect on non-agricultural employment opportunities.

The second limitation is of data. There is no time series of NDP in agriculture for individual states over the period under review. Our measure of agricultural output is therefore limited to an index of agricultural production for all crops constructed by A. V. Jose of the Centre for Development Studies Trivandrum. We have used this index to construct an index of agricultural production per head of the rural population for each state and used this as the explanatory variable measuring agricultural performance in our regressions.16 Since this is a gross output measure, which takes no account of the increased input intensity of agriculture over time, it probably overstates agricultural growth in terms of net value added.

These limitations are likely to make the empirical results at the state level somewhat weaker and this is indeed the case. There is some support for the hypothesis that the incidence of poverty is inversely related to agricultural output per head, but the corroboration is not complete. Indeed, there are interesting differences between the all-India and state-level results, which call for further investigation.

An important difference between the all-India experience and the experience for individual states is that whereas in the former case agricultural output per head of the rural population stagnated, this is not true for individual states. Table 6 presents the estimated (exponential) growth rates in IAPPH for individual states. Six states show significant growth in output per head none of these states show a significant trend decline in the incidence of poverty and West Bengal actually shows a significant trend increase. The absence of a trend decline in the incidence of poverty in states that have experienced growth in agricultural output per head is clearly disturbing and calls for further investigation. variable. The explanatory variable measuring agricultural performance is a two-year average of IAPPH where the averaging reflects the existence of lagged effects. We have also included time as an additional explanatory variable to test for the existence of underlying time.

Following the approach adopted for the all-India analysis, we have estimated regression equations for individual states testing the hypothesis that the incidence of poverty depends on the level of agricultural output per head of the rural population presents the results of two regressions for each state, using the percentage of the rural population in poverty and the Sen Index respectively as the independent trends in poverty incidence generated by factors operating independently of changes in the level of output per head. Our results can be summarised as follows.

(i) There is clear evidence of a significant negative relationship between agricultural output per head and the incidence of poverty in seven states Andhra Pradesh, Bihar, Karnataka, Madhya Pradesh, Maharashtra, Tamil Nadu, and Uttar Pradesh Although only seven States of the fourteen conform to this pattern, it is important to note that these states account for 56 percent of the rural population of India and about three-quarters of the rural population in poverty. Of the other states, it is worth noting that Orissa and West Bengal have negative coefficients on the IAPPH variable with t ratios that are fairly high, although not high enough to ensure significance at the ten per cent level for a two-tail test.

(ii) Where the state level results differ substantially from the all-India results is in the estimated coefficients on the time term. At the all-India level we found no significant time trend in poverty incidence operating independently of the effect of agricultural output per head The results for individual slates show that the coefficient on lime is significant in a number of states, and in all those cases ii is significantly positive If we accept the argument that the lime term picks up the net impact of variables excluded from our analysis, these results suggest that in these states- Assam. Bihar. Gujarat, Karnataka, Orissa, Rajasthan, Uttar Pradesh, and West Bengal-there may be factor at work in the rural economy which by themselves tend in increase the incidence of rural poverty. Identifying these factor is clearly crucial for understanding the causal mechanisms determining rural poverty. Unfortunately, our data provides no basis for developing and testing specific hypotheses along these lines. However, it is interesting to note that this group includes all the states of the eastern region where conditions of tenancy are most adverse for small farmers.

(iii) For most of the states for which the coefficient on the lime term is positive, there is also a significantly negative coefficient on the IAPPH variable. This suggests that while there were factors operating in the rural economy which tended to increase the incidence of poverty, agricultural growth leading to higher output per head tended-lo offset the adverse impact of these factors. Can we conclude from this that increases m agricultural output per head are in themselves always beneficial? The key assumption underlying this argument is that such increases can be achieved, within the existing institutional structure, without affecting the other factors which operate to increase the incidence of poverty. This is clearly a strong assumption. Some of the factors whose effect is captured by the time term obviously operate independently of any attempt to increase agricultural output. Increasing population pressure on land leading to a reduction in size of landholdings is an obvious example of such a process which is particularly relevant in rural India. However, there are other processes which may have an adverse effect on poverty and which may be speeded up if we attempt to accelerate the pace of agricultural growth These include technological change in agriculture which has a net labour displacing effect and which may therefore weaken the economic position of both landless labour and small tanners who rely in part on wage employment.

The scope for reducing the incidence of poverty through raising agricultural output per head in the future is therefore e crucially dependent upon whether this can be done without technological changes that are excessively labour displacing. The regressions reported above cannot address this question satisfactorily since the observed variation in agricultural output over the period is dominated by weather-induced fluctuations and the impact of such variations on poverty may be different from the impact of a more rapid expansion in agricultural output over time

(iv) Finally, the most disquieting feature of our results is the evidence from Punjab and Haryana which does not support the hypothesis that improved agricultural

> performance will help reduce the incidence of poverty. This region has experienced a dramatic growth in agricultural output per rural person but there is no evidence of a downtrend in the incidence of poverty. The Punjab-Haryana experience may have been wrongly described by some authors as a case of increasing incidence of poverty despite rapid agricultural growth, when the evidence shows no significant time trend, but it is scant reward for the most successful agricultural performance of any state if the poorest 25 per cent of the rural population experienced stagnant levels of real consumption.

How do we account for this outcome? One approach is to attribute it to the particular nature of the agricultural growth experienced in this region. Stagnation in real consumption levels at the lower end of the scale could arise from strongly labour-displacing technological changes accompanying agricultural expansion, exemplifying the malign effects of some types of growth. However, there is also a benign explanation, attributing the lack of any reduction in poverty incidence to the fact that there has been heavy migration into rural Punjab in response to the growth of labour demand. It can be argued that the poorest quarter of the rural population in Punjab and Haryana contains an increasing proportion of in-migrants, and although real consumption levels of the poorest quarter appears stagnant over time, this group consists increasingly of individuals whose consumption is higher than it would have been if they had not migrated. In other words, 'trickle-down' benefits have taken the form of increased employment benefiting migrants from other states, rather than increased wages benefiting the pre-existing poor. Such supplementary evidence as is available certainly points to a substantial growth in labour demand and in-migration into the region together with an increase in average real wages, developments which are difficult to reconcile with the malign view of the impact of agricultural growth.

A further aspect of the Punjab results which must be kept in mind is that poverty in this region is closer to being a problem of low end poverty than in other states. As such, the insignificant coefficients on the IAPPH variable may arise from the fact that even if trickle-down processes do exist in the rural economy, they may not reach all the way down to the very poor. It has long been recognised that for such groups, poverty alleviation will require special programmes of assistance and support and not merely a general improvement in productivity. The evidence provides some support for this view If we redefine the poverty line for this region to be 20 per cent higher, and use the resulting estimates of the incidence of poverty as dependent variables in our regression equations, we find a statistical improvement over the results. When the dependent variable is taken as the percentage poor, we find a positive coefficient on time and negative coefficient on the IAPPH variable with both coefficients significant at the five per cent level. When the Sen Index is used, the coefficients have the same sign but they remain insignificant at the ten per cent level.

The Evidence on Relative Inequality

Thus, far, the distributional implications of increases in agricultural output per head have been examined solely in terms of the impact on absolute poverty. It is also useful to examine the evidence on relative inequality in consumption (and changes in inequality over time) in order to determine whether it supports the benign or malign view of the distributional impact of agricultural growth in India.

The view that agricultural growth within the present institutional constraints does not contribute to poverty alleviation, and indeed may even generate absolute impoverishment for the poor in the sense of declining real incomes, implies that we should also see increased relative inequality in the distribution of consumption. This would be the case as long as the consumption function is monotonic, since income increases for the non-poor would be reflected in increased consumption for these groups while stagnant (or reduced incomes for the poor will lead to stagnant or lowered consumption levels.

The NSS data provide the basis for documenting trends in the inequality of consumption expenditure in nominal terms Ideally, we should examine trends in the distribution of real consumption since changes in the distribution of nominal consumption may reflect no more than differential movements m the price indices However, in the absence of fractile specific price indices we can only compare inequality in nominal terms Such comparisons remain of interest since the available evidence suggests that while they may exaggerate the extent of the change m inequality, they nevertheless point in the right direction presents the Cum coefficients of the distribution of rural consumption expenditure for India as a whole and for the individual states We have tested for the existence of a linear time trend in each case and the results are presented in Table 9 from finding an increase in relative inequality, we find that the evidence points in the opposite direction. I here is a significant decrease in relative inequality for India as a whole and for seven of the fourteen stales Of the other stales, all except Kerala have a negative coefficient on the time term, although none of' these coefficients is statistically significant It is particularly worth noting that the Gini coefficient for Punjab and Haryana shows a statistically significant decline in inequality over the period as a whole.

The absence of any marked increase in relative inequality as measured by the NSS is vet another reason for questioning the view that agricultural growth has been accompanied by absolute impoverishment for the rural poor. As shown in Table 6, six states Kerala, Orissa, Punjab and Haryana, Tamil Nadu, Uttar Pradesh and West Bengal show a significant trend increase in output per head of the rural population. We note that none of these states experienced the increase in relative inequality which should have occurred if growth was accompanied by absolute impoverishment. Indeed three of them show a significant decline in consumption inequality.

The evidence reviewed provides a fairly firm basis for documenting trends in rural poverty in India. We find that the Indian experience over the past two decades cannot be characterised as showing a trend increase in the incidence of poverty in India as a whole. The same conclusion holds for all the individual states except Assam and Bengal, which show a significant trend increase in poverty. In general, the time series shows a pattern of fluctuation, with the incidence of poverty falling in periods of good agricultural performance and rising in periods of poor performance. Given the importance of weather-induced variations in Indian agriculture, there can be little doubt about the importance of such fluctuations and it is crucial to keep these in mind in assessing underlying trends.

The evidence on the relationship between rural poverty and agricultural performance is more difficult to evaluate for two reasons. In the first place, the evidence itself is somewhat mixed. Much depends upon the level of aggregation at which the analysis if conducted with the all-India results presenting a somewhat different picture from that obtained at the level of individual states. Furthermore, the evidence necessarily is difficult to interpret since we are implicitly searching for causal relationships in what are at best observed correlations. The all- India evidence is entirely consistent with the hypothesis that the incidence of rural poverty is inversely related to agricultural performance measured in terms of agricultural NDP per rural person. Taken at face value, this correlation suggests that faster agricultural growth, by raising agricultural NDP per rural person, might have led to a reduced incidence of poverty The absence of any adverse underlying time trends further supports this view.

Chapter 2

Can Non-Agricultural Employment Reduce Rural Poverty

Several recent studies have addressed the importance of off-farm incomes in rural areas of the developing world. Evidence from El Salvador, Ecuador, and Mexico suggests they constitute between 40 to 50% of total rural households' income, 30-50% in Sub Saharan Africa, and up to 80% in South Africa. Similarly, data from India in the early 90s shows that off-farm incomes represent 34% of total households' income. Moreover, the Indian data shows that the importance of off-farm incomes over total income does not vary dramatically across income quintiles.

The range of activities classified as non-agricultural jobs is broad. For example, in Northeastern Brazil where 22% of the rural working population participates in non-agricultural activities as their primary occupation, of those in non-agricultural jobs, one third is in the manufacturing sector, 17% in commerce, 13% are self-employed, and 18% in education and government. The heterogeneity in the types of occupations suggests that the non-agricultural sector encompasses wide variation in terms of wages, working conditions, stability, and contracts. Along these dimensions, there can be differences in the access of the poor to non-agricultural jobs. For example, the India study found that casual employment was more prevalent among the lower income quintiles, while permanent employment was more common among the higher quintiles.

Despite the increasing importance of off-farm incomes in rural areas, it has been difficult to address empirically what is the impact of non-agricultural rural employment on poverty reduction. In the policy world, many have pointed towards the potential of off-farm employment to reduce rural poverty. However, there are reasons to believe the poor may indeed not be the first ones to benefit from NARE. For instance, several studies that look at the individual probability of participation in the sector find that it is usually the more educated young individuals, with better social networks and access to roads, who are more likely to participate in this sector. This suggests that access to off-farm jobs may be limited for the poor. Similarly, in contexts where land and other assets are concentrated, spillovers from linkages between agriculture and the non-farm sector may not reach the poorest. Previous work has documented that inequality can affect not only growth, but its poverty reduction potential and as the prospects of fast growth of agriculture in the developing world on the near future are small, the poverty reduction potential and distributional implications of NARE deserve a closer look.

This paper focuses on two questions. The first question relates to the impact of NARE on poverty. The second one explores in which environments can policy interventions in education and roads be more effective in reducing poverty through NARE.

These questions are addressed empirically using Mexican municipal data for the 1990s and through the estimation of employment, poverty, and population equations. Results show a negative effect of manufacturing and services employment on poverty. The effect of manufacturing employment is larger than the one of services in semi urban municipalities. In turn, services employment has a larger effect on poverty in rural than in semi urban municipalities. When exploring the effectiveness of policy interventions in reducing poverty through NARE, the data suggests that interventions to improve roads and secondary education reduce poverty through services employment in rural and semi urban municipalities and through manufacturing employment only in semi urban municipalities. While these interventions are only poverty-reducing through manufacturing employment in semi urban municipalities, they are more poverty-reducing through services employment in rural than in semi urban municipalities.

Background

Some of the related literature on the linkages between NARE and poverty. In understanding this relationship, many have focused in the differences across the off-farm jobs of the poor and those of the non-poor. For example, in their study of Northeastern Brazil, Ferreira and Lanjouw find that rural areas near urban municipalities account for a much larger share of non-farm employment than isolated areas. This evidence is consistent with the importance of access to markets and infrastructure for the expansion NARE. Interestingly, when looking at the importance of income from NARE over total income, they find that its share is homogeneous across income quintiles. The differences lie across other dimensions of off-farm occupations. For example, while cultivation income is more important for those in higher income quintiles, agricultural labor plays a larger role for the lower quintiles. They also find that non-farm enterprise income is more important for those with larger incomes. On the contrary, incomes from low return off-farm activities are only important for the landless. These patterns, and the large prevalence of occupations in sectors such as construction and domestic services among the off-farm jobs to which the poor have access, suggest that some of NARE consists of residual occupations to which people who do not have other options are driven.

Along this line, when trying to disentangle the link between economic growth and poverty, Lanjouw and Shariff characterize the off-farm sector as one where two types of occupations are prevalent: some that are productive such as manufacturing and services and other that have low productivity. The latter contribute to the incomes of the poor, but do not allow for social or economic mobility. The authors suggest that because of the presence of these two types of occupations, it is not clear in which direction the importance of off-farm income evolves with economic growth. However, they suggest there may exist an indirect impact of the off-farm sector on poverty through wages, as the expansion of the non-agricultural sector can push agricultural wages up.

There are many parallels that can be found between the literature that looks at the poverty reduction potential of NARE and the studies on pro-poor economic growth. Among the cross-country work in this area, Timmer explores the "elasticity of connection" between economic growth and growth in per capita income. He finds that elasticities are not equal to one and that they rise with income. This suggests that economic growth can worsen the distribution of income. Timmer finds that in countries where the income gap between the rich and the poor is small, labor productivity in agriculture is more important in generating incomes for all quintiles. Also, agricultural productivity results in higher elasticities of connection for the lower quintiles than for the higher ones. Finally, where the income gap is small, he finds that growth of the non-agricultural sector reaches the poor.

Ravallion and Datt argue that while cross-country evidence shows poverty falls with economic growth, there is variation in the povertyreduction potential of a given rate of growth. Using Indian data, they explain whether differences in poverty reduction are due to heterogeneity in rates or sectorial patterns of growth or in the poverty reduction potential of growth. They find that farm yields, government expenditures, and non-farm output are associated to reductions in poverty, and that inflation is poverty-increasing. Their results suggest that NARE growth is more pro-poor in places with higher farm yields and female literacy, and with lower infant mortality, urban-rural disparities in consumption, and landlessness. In their study of India, Ravallion and Datt conclude that the sectorial composition of growth matters to the aggregate rate of poverty reduction. While initial conditions in rural development and human capital account for much of the differences in rural poverty reduction, they enter in an additively fashion.

The previous discussion of the literature on the relationship between poverty and NARE points towards several elements that can influence its effectiveness in reducing rural poverty, such as the sectorial composition of NARE, local inequality along various dimensions e.g. income, land, or education labor productivity in agriculture, agricultural potential, and government expenditures. The analysis which follows incorporates all of them.

Model

The proposed framework explores the relationship between NARE and poverty, and the role of the local context on the effectiveness of policy interventions to reduce poverty through the expansion of NARE. It starts with an analytical model that relates employment and poverty reduction. The model consists of three key components: an equation of employment, one of population changes, and a third one of poverty.

A second subject of interest are the intersectorial linkages in terms of employment. In the presence of positive employment linkages, one would expect to find that in places where employment in non-agricultural activities grew faster, employment in agriculture also experienced a bigger expansion explores these correlations through a smoothing of non-agricultural on agricultural employment growth. The first clear pattern is that along the region of negative agricultural employment growth, there is positive non-agricultural employment growth for both manufacturing and services, or substitution between agricultural and non-agricultural employment. In the positive region of agricultural employment growth, there are two trends. For all of the curves but manufacture in semi urban municipalities, the same pattern of substitution remains. However, for manufacture in semi urban municipalities is consistent with the presence of positive externalities from agriculture in terms of NARE generation. explores the correlation between manufacturing and services employment growth, again by smoothing one on the other. In this case, the patterns are suggestive of positive employment linkages, or complementarity. This positive correlation between manufacturing and services employment growth is larger in magnitude in semi urban than in rural municipalities.

Estimation

By all three measures, poverty is higher in rural than in semi urban municipalities, while the level of employment in manufacturing and services is larger in the latter. Population between 1980 and 2000 grew at a much faster pace in semi urban than it did in rural municipalities. While income inequality seems larger in rural than in semi urban municipalities, there are few differences across them in terms of the two agricultural variables. However, the per capita municipal expenditure was larger in rural than in semi urban municipalities.

The geographic variables show that more semi urban than rural municipalities are in the coast, and more rural than semi urban municipalities are located in the high-plateau of central Mexico and in areas where the terrain is more irregular. Similarly, rural municipalities have a larger indigenous population and are more isolated from an urban center than semi urban ones are.

Finally, the other municipal variables show there are important differences in terms of schooling, as semi urban municipalities have, on average, a percentage of adults with some secondary education that is twice as high as that of rural municipalities. Semi urban municipalities are also better served by state roads and are more likely to be among the municipalities for which the law mandates high minimum wages. While there is little difference in terms of the average wage in manufacturing across semi urban and rural municipalities, wages in services are much higher in the former than they are in the latter present the results of the estimations. Separate models were computed for rural and semi urban municipalities and for employment in services and in manufacturing. In all of the estimations, the explanatory variables were demeaned so the parameter in front of an interacted variable can be interpreted as the total marginal effect of the interaction.

The third step of the estimation incorporates the predicted values of per capita manufacturing and services employment and of population to estimate poverty in 2000. T-tests compare these coefficients across rural and semi urban municipalities. Sargan tests do not reject the validity of the instruments although for semi urban municipalities only at the 5% level. The first two rows of the table refer to the total effects of per capita manufacturing and services employment on poverty. There is a negative association between poverty and NARE in all cases, except manufacturing in rural municipalities, where the coefficient is not significant. In semi urban municipalities, the negative effect of manufacturing employment on poverty is larger than the one of services employment. Also, while manufacturing employment is poverty-reducing only in semi urban municipalities, the poverty-reducing effect of services employment is larger in rural than in semi urban municipalities.

Effectiveness of Policy Interventions

On the other hand, the effect of interventions in roads is poverty-reducing through both manufacturing and services employment in semi urban municipalities and through services employment in semi urban municipalities only. Just like with interventions in education, the effectiveness of roads in reducing poverty through services employment is also larger in rural than in semi urban municipalities.

From a policy perspective, it is interesting to explore whether heterogeneity along different dimensions may result in differences in the effectiveness of these policies in reducing poverty through employment in manufacturing and services.

However, these interventions are more effective in municipalities where income inequality and the value of agricultural output are high than in those where they are low. Also, interventions in education are more effective in reducing poverty in semi urban municipalities where the labor productivity of agriculture and the municipal government expenditures are low than in places where these are high. The effectiveness of roads in reducing poverty in semi urban municipalities through the promotion of NARE suggests the opposite pattern. Here, interventions are more effective in places where income inequality and the value of agricultural output are low. Also, roads are more effective in reducing poverty through NARE in semi urban municipalities where the labor productivity of agriculture and municipal government expenditures are high than in those where they are low.

This exercise is more insightful among rural municipalities, where it was found that the total effect of policy interventions on poverty through manufacturing employment was not poverty-reducing allows us to distinguish contexts in which it is. For instance, the total effect of interventions in secondary education is always poverty-reducing. Moreover, its effectiveness in reducing poverty through NARE is larger in rural municipalities with high value of agricultural output than in those with low. This is also true in municipalities where income inequality, labor productivity of agriculture, and government expenditures are low.

Conclusion

This paper answers two questions. First, it assesses the impact of nonagricultural rural employment on poverty. And second, it explores in which environments can policy interventions in education and roads be more effective in reducing poverty through NARE. In understanding the poverty-reduction potential of NARE, it explores the role of other municipal attributes such as the sectorial composition of NARE, local income inequality, labor productivity in agriculture, agricultural potential, and government expenditures. The findings suggest that there is a negative effect of manufacturing and services employment on poverty in semi urban municipalities and a negative effect of services employment on poverty in rural municipalities. The effect of manufacturing employment on poverty in rural municipalities is not significantly different from zero. This poverty-reducing effect is larger for manufacturing than for services employment in semi urban municipalities. Also, the negative effect of services employment is larger in rural than in semi urban municipalities. Our findings suggest that poverty is higher in municipalities with higher income inequality, and with lower government expenditures. These two effects are stronger in semi urban than in rural municipalities. Also, higher values of agricultural output are negatively related to poverty and this effect is larger in rural than in semi urban municipalities.

The last part of the paper compares the effectiveness of interventions that increase the availability of secondary education and of roads in reducing poverty through the expansion of NARE. Interventions in secondary education are more effective in reducing poverty through NARE in places with low labor productivity of agriculture and low per-capita government expenditures and in places with high value of agricultural output. However, the effectiveness of secondary education in reducing poverty through NARE differs across rural and semi urban municipalities in terms of inequality. While these interventions are more poverty-reducing in rural municipalities with low income inequality, the opposite is true for semi urban municipalities.

Finally, interventions in roads are more effective in reducing poverty through NARE in semi urban and rural municipalities with low value of agricultural output, but high labor productivity of agriculture. The effectiveness of these interventions does not differ for rural municipalities along dimensions such as inequality and municipal government expenditures. However, for semi urban municipalities, interventions in roads are more effective where income inequality is low and government expenditures high.

Chapter 3

Rural Development and Poverty Reduction: Is Agriculture Still the Key

This paper examines the relationship between rurality and poverty, and the role the agricultural sector can play in rural development, poverty reduction, and overall development. The historical views regarding the role of the primary sector in development are presented, and then using original data, the paper argues that there was an historical misjudgment against the primary sector that served as a foundation for anti-agricultural bias in public policy until the late 80's. Finally, this paper explains how under certain conditions territorial/regional development strategies may prosper, but in other conditions, particularly in the least-developed countries rural space, agriculture is still necessarily the starting point for rural development. The designations employed and the presentation of material in this information product do not imply the expression of any opinion whatsoever of the part of the Food and Agriculture Organization of the United Nations concerning the legal status of any country, territory, city or area or of its authorities, or concerning the delimitation of its frontiers or boundaries.

Introduction

It is not an exaggeration to say that the battle to achieve the global society's stated objectives on hunger and poverty reduction will be won or lost in the rural areas of the developing countries. Globally, extreme poverty continues to be a rural phenomenon despite increasing urbanization. Of the world's 1.2 billion extremely poor people, 75 percent live in rural areas and for the most part they depend on agriculture, forestry, fisheries and related activities for survival. The promotion of the rural economy in a sustainable way has the potential of increasing employment opportunities in rural areas, reducing regional income disparities, stemming premature rural-urban migration, and ultimately reducing poverty at its very source. In addition, development of rural areas may contribute to the preservation of the rural landscape, the protection of indigenous cultures and traditions while rural societies could serve as a social buffer for the urban poor in periods of economic crisis or social urban unrest.

However, public policies at national level and resource mobilization at both national and international levels have not always recognized the multiple potential of the rural economy. Public policies and investments in developing countries have historically favored industrial, urban and service sectors at the expense of agricultural and other rural sector development. In many cases, a coherent rural development policy has fallen victim of the lack of a cross-sectoral institutional framework.

The past 20 years have witnessed a steep decline in the availability of public resources for agriculture and rural development. Between 1983-1987 and 1998-2000, the annual average allocations of Official Development Assistance for agriculture in the leastdeveloped and other low-income countries fell by 57 percent from USD 5.14 billion to USD 2.22 billion. Lending from international financial institutions followed a similar pattern while domestic public spending has remained stagnant at best. The result has been reduced incentives for rural investment. Serious questions have also been raised as to the efficiency and effectiveness of public resource mobilization for agriculture and the rural space. For example, López and Galinato show a consistent bias in rural spending in Latin America in favor of subsidies and against investment in public goods which has translated into lower agricultural growth in the region.

An important question in the development debate regarding rural development has been the relationship between agriculture and the rural economy. In certain respects, past policy perceptions and practice have often equated rural development with agriculture, and rural development policies have been subsumed under an agricultural policy package. The issue of how and under what conditions agriculture is a driving force of rural growth has received scant attention or has given mixed messages including in the position of major multilateral financing institutions.

Recently however, the emergence of national and international commitments on poverty and related targets as for instance in the Millennium Development Goals and the Poverty Reduction Strategies at country level coupled with the failure of past paradigms to make mass reductions in rural poverty, have given a new impetus to the role of agriculture in development and poverty reduction. While at the same time, new rural development models have emerged especially in the context of Latin America emphasizing a more broad approach in which rural and urban space are viewed as a continuum and their interactions are emphasized.

The principal objective of this paper is to present in a concise way, some of the conceptual issues regarding the role of agriculture in both rural and overall development, emphasizing its role in poverty reduction; and to explore if agricultural development can be an engine of growth and poverty reduction in developing countries, and under what conditions.

The paper is structured as follows: In the next section we deal with issues of definition of rurality and rural development. Next, we explore the link between poverty and rurality. Then we discuss the role of agriculture in development, rural development and poverty reduction. We then discuss contexts in which spatial constraints make a sectoral development entry point necessary. Finally some concluding remarks are provided.

What is Rural Development

The definition of rural development has evolved through time as a result of changes in the perceived mechanisms and or goals of development. A reasonable definition of rural development would be: development that benefits rural populations; where development is understood as the *sustained* improvement of the population's standards of living or welfare.

This definition of rural development, however, has to be further qualified. In the 1960's and early 1970's the consensus was that intense industrialization was the main characteristic of the perceived development path. In this context it seemed natural to define rural development as precisely leading into that path: "Rural development is essentially a part of structural transformation characterized by diversification of the economy away from agriculture. This process is facilitated by rapid agricultural growth, at least initially, but leads ultimately to a significant decline in the share of agriculture to total employment and output and in the proportion of rural population to total population." Later during the 70's, mostly based on equity considerations, the focus and definition of rural development turned to the provision of social services to the rural poor. This shift was partially founded on the recognition that even under rapid growth of income in rural areas, the availability or equitable access to social services and amenities was not guaranteed.

Lacroix exemplifies this line of thought when he explains the difference between agricultural and rural development: "Agricultural Development generally tries to raise agricultural production and productivity and is of a technical nature. It is similar to other efforts to develop physical capital as a means for economic growth. Rural Development, though, by definition is oriented more toward benefiting primarily the poor. Thus, the fundamental distinction between pure agricultural and rural development is the emphasis on capital development for the former, and human capital development for the latter."

Since the 1970's rural development as a concept has been highly associated with the promotion of standards of living and as a precondition for reducing rural poverty. This propoor bias was born from the understanding that, particularly in societies where wealth is extremely concentrated, mean incomes could grow without improving the well being of the most dispossessed. Thus, if the general definition of rural development is accepted, i.e. the improvement of the welfare of *all* members of the rural populations, then this pro-poor bias is justified.

On the other hand, the focus on human capital formation, through the provision of social services in rural areas has been constantly stressed since the 1970's. Originally, this focus stemmed from social equity considerations: it is fair that all of society's members have access to services like education and health. However, the development of endogenous growth theory in the late 1980's provided macro-foundations for this priority, as this theory proved how permanent growth development is possible even in the presence of constant returns to scale when there is balanced investment in both human and physical capital at the same time.

What is Rural

Having defined rural development it is essential to define what is rural. Unfortunately there does not exist a single methodology, much less a single definition of what constitutes rural. The problem is that patterns of spatial occupation are, *inter alia,* culturally and historically determined and vary among regions of the world. A natural definition of rurality is to define it by exclusion, as that which is not urban, where urban is defined on the basis of population agglomerations. In practice there are two main methodologies to define rural. The first methodology is to use a geopolitical definition. First, urban is defined by law as all of the state, region, and district capitals, and by exclusion all the rest is defined as rural. Countries like Colombia, El Salvador, Dominican Republic, and Paraguay follow this methodology. In all of these countries urban population is defined as that living within the "*cabecera municipal*" the municipality's head or center. The drawbacks of this methodology are obvious populations that live outside the geopolitical limit of a city specially in a growing city are miscounted as rural; while population living in tiny municipalities in sparsely populated regions is miscounted as urban.

The other popular methodology is to use observed population agglomeration to define urban. In this case populations that live within an area where contiguous households form populations larger than, say 2,000 inhabitants are considered urban, while by exclusion the rest is defined as rural. This methodology seems more attractive because it establishes a clear threshold; unfortunately this threshold varies widely around the world. In countries like Uganda, an agglomeration of only 100 inhabitants constitutes an urban settlement, while in countries like Nigeria and Mauritius the minimum agglomeration for urban areas is 20,000, and even 30,000 in Japan. In spite of these large variations certain thresholds are popular, like 2,000 in Chile, Argentina, Bolivia, Israel, and France; 2,500 Mexico and USA; and 5,000 in Belgium and Switzerland.

Other less popular methodologies consist of counting agglomerations of homes; for example, 100 contiguous dwellings constitutes an urban area in Peru. Other countries consider the availability of services as defining urban. For example in Honduras, an area is urban if, in addition to having a population of 2,000 inhabitants, it possesses services of water, electricity, education and health infrastructure. This definition is relevant from a poverty analysis standpoint because the absence of these services is usually associated with poverty; thus, this particular definition of rurality provides elements for characterizing poverty. In comparison, rural areas in developed nations have availability of all of the above mentioned services, spatially more scattered, but still available.

The limitations in comparability imposed by the differing definitions of rurality have spurred efforts to create internationally comparable measures of rurality. These projects merge satellite imagery, which shows population agglomerations, together with census data to spatially distribute populations. The GRUMP project is one major effort of the type, identifying cities with night-light satellite imagery and using census data to distribute population inside and outside those cities4. Although the effort is important, it is still not an ideal measure because it depends on the reliability of the original census and its degree of sub-administrative unit detail, and because it uses official and therefore different rurality figures in the model used to assign rural and urban populations.

For Latin America, Chomnitz et al. created a rurality indicator based on population densities and distance to a major city. The cutoff point to define rurality is areas with populations living in densities below 150 inhabitants per squared kilometer *and* living more than 1 hour of travel away from a major city. We summarize these different measures of rurality. The table shows that there is less variation of rurality in the internationally comparable figures; but more importantly, it also shows that rurality is under-reported in official figures. These underreporting is more pronounced in Latin America where GRUMP-measured rurality is about 10% higher than the official numbers. The averages weighted by population, in parenthesis in the tables, show that the underreporting is even higher in Sub Saharan Africa and in Latin America.

Although it seems easy to think of what is urban and rural, the multiplicity of definitions shows that it is not as easy to define. Furthermore, the fact that the definitions vary so greatly, creates a problem for making meaningful comparisons, when "rural" refers to a variety of different contexts. Also, if "rural" characterizes a variety of contexts, it is hard to see how one single rural development strategy even in the same country may be applied. Finally, the beneficiaries of a successful rural development strategy, the rural populations could be larger than what official figures indicate.

Rurality and Poverty

Most of the world's poor live in rural areas. IFAD estimated in 2001 that among the poorest 1.2 billion people in the world, surviving with less than a dollar per day, three out of four lived in rural areas5. They constitute the poorest fifth of world population and do not earn enough to cover their food needs. In a recent World Bank study, Ravallion et al. estimate that in 2002, 75% of the developing world poor still live in rural areas. Part of this correlation between rurality and poverty is given by the fact that some countries, as we have seen, indirectly define the poor as rural. As one of the most accepted characteristics of development is a secular decline in the share of agriculture, countries with larger rural populations shares are expected to be poorer since the main activity in the rural economies is likely to be agriculture.

However, the question begs an answer, how much are poverty and rurality correlated. To get a measure of correlation we collect all the available figures of poverty measured as the percentage of population in each country living under 2 Purchasing Power Parity US dollars per day. The benefit of this measure of poverty is that it is comparable, because it offers a comparable poverty line which accounts for the differences of purchasing power of 1 dollar in each different country. Recognizing the limitations associated with the measure, we regressed the percentage of rural population against poverty. We are obviously not surprised to find a positive, significant and high correlation. The coefficient of 0.83 translates to an elasticity of 0.86; that is, for each percentage point of reduction in the relative share of rural population, poverty is reduced by 0.9 percentage points. What is truly surprising is to find such a high fit for this simple regression. About 85% of the differences in poverty rates are explained by the rurality level alone. The surprise should be compounded if one considers the differences in the definition of rurality between countries.

The Importance of Agriculture in Development

In this section we provide a brief historical account of some of the most significant views about the role of agriculture in development.

Economist as early as the beginning of the 20th century observed that wealthier countries were characterized by a smaller portion of their output coming from agriculture and relatively less labor resources tied to the same sector. They also noted that the process of development itself was characterized by a monotonic decline in the relative importance of agriculture and the primary sector in the economy, both in terms of GDP and employment. One of the first economists to point this out was G. B. Fisher. Later, this same generalization was formalized by Kuznets, showing that this secular decline of the primary sector with development can be observed both across countries and across time.

Therefore, if the process of development is characterized by a shrinking agricultural sector, should the development "recipe" then suggest policies that are biased against agriculture in favor of other sectors of the economy to accelerate development? Or should agricultural growth be promoted to facilitate this structural transformation? If one looks at history, and recounts the policies that developing countries implemented from the late 1950s until the 1980s particularly in Africa and Latin America, it would seem that they followed the first strategy.

Some economists argued that agriculture plays an important role in development. One of the first arguments in favor of the role of agriculture in development was placed by Lewis who suggested that "there are large sectors of the economy where the marginal productivity of labor is negligible, zero, or even negative." Of course these labor resources are tied to the primary sector, and are a key ingredient for industrial growth, which will occur thanks to a growing labor force coming from the primary sector. Hence, the primary sector plays, although passive, an important role in development. Later, Johnston and Mellor, identified some active roles that the agricultural sector performs throughout the development path: (*i*) agriculture provides food necessary for a growing economy, as food demand, although at a decreasing rate, grows with income (*ii*) agricultural exports generate the foreign exchange necessary to import capital goods; (*iii*) agriculture, as the larger sector in less developed countries, is the only sector capable of generating the savings mass that the nonagricultural sector needs for capital accumulation; and (*iv*) a growing agricultural sector creates a larger local market for the non-agricultural sector. These Johnston-Mellor linkages still remain relevant for developing economies with a large primary sector.

Johnston and Mellor were perhaps also the first to note that successful industrialization experiences are usually preceded by periods of dynamic agricultural growth. Although this does not amount to a causality link, the authors observed that countries that embark in a successful industrialization path, first experience fast agricultural expansion, fueled not by absorbing resources from the rest of the economy, but by rapid increases in productivity. The authors tell the story of Japan in the early 20th century. Many others have mentioned this feature of development; today for example, we can tell a similar story for the Asian giants of India and China, with fast industrialization preceded by fast productivity growth in the agricultural sector, i.e. the "green revolution". As a matter of fact, countries that have been able to industrialize without having first an agricultural expansion are the exception.

However this was not a widely accepted message. During the second half of the 20th century there was no widely shared optimism concerning the role of agriculture in development. Agriculture was in practice condemned mainly by two separate schools of thought. The *structuralist* school particularly strong in Latin America provided arguments against agriculture. What eventually became to be known as the Prebisch-Singer hypothesis states that the commodities that developing countries the periphery produce and export have an income elasticity of demand less than one; as opposed to the demand elasticity of the industrial goods produced by the developed countries that have income demand elasticity that is not less than unity. Therefore in the long run the price of the primary commodities exported by developing countries relative to the price of the industrial goods imported by these same countries is doomed to fall. Thus, specializing and exporting these primary commodities is a losing development proposition.

This hypothesis was disseminated before reliable price time series were available. Today with good statistics it is a contentious debate if this hypothesis holds or not; principally because it depends on the primary commodity and the industrial manufactures price index used as deflator. Also, should one account for changes in quality? Recent long-term analysis suggest that for most commodities the hypothesis does not hold.

In the short-run most commodity prices are pro-cyclical. In the long-run some commodity real prices are non-stationary and thus move around a stable mean, while other commodities show one time falls in early 1920s and 1980s, and some other commodities show an upward trend like the relative price of meat. Therefore, the Prebisch-Singer hypothesis may apply to explain the failed development experience of any one particular country, dependent on a particular commodity, but it can not be generalized to all developing countries and to all commodities.

The second main hypothesis that condemned agriculture is related to the belief that agriculture has low potential for growth and its multiplier effect, its ability to "pull" the rest of the economy, is very low. The first part of this hypothesis can be traced back to Rosenstein- Rodan. He proposed that industrialization can be achieved by investing in several different industries separately, even if none of them *alòne* generates enough rents to break even. This is possible because there exist economies external to the industry that spill-over to the rest of the economy. If there are enough industries generating this type of positive externalities, the "big push" into industrialization can be generated. Agriculture here lies in its absence, it is a sector that does not generate these economies external to the industry.

Also part of this agro-pessimism, and perhaps more influential in condemning the sector is Hirschman's work *The Strategy of Economic Development*. Hirschman opposed what he called a "balanced growth" strategy, of all sectors growing at the same rate, i.e. Rosenstein-Rodan "big-push" argument. Hirschman advised promoting the growth of the sector with the greater ability to pull the rest of the economy. He focused on the production backward linkages, that is the links in production that one sector has with the rest of the economy as a purchaser of inputs. If a sector A with high backward linkages expands, the rest of the economy will consequently experience a larger expansion, as it sells the inputs that sector A needs to grow. To uncover the backward linkages, Hirschman analyzed the inputoutput matrices of countries with available data: Italy, United States and Japan note that all are developed countries, and this is not inconsequential as we show below, and discovered that agriculture has high forward linkages, but among the lowest backward linkages of any other sector. "Agriculture argues Hirschman certainly stands convicted on the count of its lack of direct stimulus to the setting up of new activities through linkage effects the superiority of manufacturing in this respect is crushing". This indictment, as we show below, examined using recent data turns out to be a historical misjudgment.

Below, evidence is presented to show that neither of these two propositions is really corroborated by the data. If industries are inherently superior to agriculture, they would manifest a long-term productivity growth rate higher than agriculture. This appears not to be the case. Faruqui et al. for example, estimate the annual labor productivity growth of the primary industries in the USA at 3.1% while that of the manufacturing sector at 3.3% per year. A simple observation of the development process, in which labor employed in the agricultural sector falls, while output still grows, shows how there is a considerable increase in labor productivity in the primary sector during development. This simple observation is corroborated in studies like Parry that show high labor productivity growth in resource industries. Of course, one can find examples of the contrary, low or even negative productivity growth in primary industries, but this is by no means an inherent characteristic of the sector.

The other "count" of low backward linkages, also does not hold as we proceed to show.

Towards a New Perspective of the Role of Agriculture in Development

Agriculture's Production Linkages and Development

Linkages in the agricultural sector are easy to identify: forward linkages are mainly in the agricultural and food processing industries, in the service industry with the restaurant and hotel industries and, sometimes, public schooling. Similarly the main backward linkages are with the agricultural industries that produce animal feed, with the chemical and mineral industry for purchased fertilizers, and, depending on the degree of sophistication of the agricultural sector with the financial and business services sector and the industry of the economy by serving as an input. The backward linkages refer to the connections of a sector as a purchaser of inputs from the rest of the economy. machinery manufacture. In many developing countries, backward industries consist of a large number of small firms fertilizer mixing, small scale transport, agricultural implement repair, commerce, etc. largely labor intensive and vital for the rural economy.

Common wisdom, as reflected in Hirschman's indictment of agriculture, suggests that as the agricultural sector becomes more developed, its backward linkages increase, by requiring more financial services, machinery and other purchased inputs.

Also, the forward linkages are more important in a developed economy, where there is an existing and more developed food industry, and equivalently a hotel and restaurant industry. Therefore, common wisdom suggests that in terms of linkages, agriculture would not be a good sector to promote in early stages of development, because its linkages and thus its multiplying effect is low. These are important questions of rural and overall development that have rarely been studied systematically.

Testing the " linkages" hypothesis

The place where the linkages between sectors are recorded in an economy is the inputoutput matrix, which shows how the total output of each sector is distributed between final consumption, from households and the government, and intermediate inputs sales; describing how each sector sells inputs to all the other sectors of the economy. In the 1950's, when there was more faith in planning, these I-O matrices where used to discover the sector with the higher backward linkages, that is the sector that could have a greater effect in "pulling" the rest of the economy. The method applied was to calculate and find the higher Leontief multiplier. When a sector expands it demands inputs from other sectors to grow; in turn, these other sectors to supply these inputs need to expand and demand more inputs from yet some other sectors. This process continues infinitely, but in each round the size of the expansion is smaller. The limit of this process of expansions generated by the growth in one sector is given by inverting a transformation of the I-O matrix, which solves for the Leontief multipliers. This method has been usually criticized because as inputs are calculated in values, and not in units, any I-O analysis assumes fixed relative prices, an unrealistic assumption when the relative size of sectors change significantly. This criticism is valid, but the I-O analysis still provides relationships that are valid at the margin.

In spite of the mathematical beauty of the inverted Leontief multiplier matrix, the problem with the multiplier analysis is that the second, third and *n*th round of backward linkage caused expansions, although progressively smaller, may not realize; either because of frictions in the economy, or because these rounds take time to complete, and by the time they do, the economy may have changed. Furthermore, this multiplier analysis does not take into account the relative importance of sectors in the economy. For example, the sectors "multiplied" by another sector may be relatively insignificant for the economy in terms of income generated.

To deal with these drawbacks of the multiplier analysis, Anríquez et al. proposed an alternative method to calculate both forward and backward linkages.

Equation (2) shows that the portion of the value added of sector *j* that is considered as a forward linkage is the ratio of inputs from sector *k* to total inputs used in sector *j*. For the purpose of this study we have collected as many I-O matrices as we could find to measure forward and backward linkages of the agricultural sector and to study any relations they may have with development. Table 4 presents the results for all the 26 countries we could find public information. As countries have been sorted by per capita GDP, the table shows the well known fact that the share of agriculture in total GDP tends to fall as countries get richer. What happens with the value of linkages, on the other hand is not known, and rather surprising. Backward linkages, contradicting "common wisdom" seem to be much higher at *earlier* stages of development and not later. On the other hand, forward linkages seem to be larger at medium income levels in our sample.

The fact that the value of linkages drops at higher levels of income makes sense, because the share of agriculture in total income is falling, and so does the value of the connections with the rest of the economy. Simple regression analysis, as demonstrate that this correlation is statistically significant. That is, both forward and backward linkages are higher for countries at earlier stages of development note that no causality is implied Thus as countries are in their earlier stages of development, agriculture has a higher effect in national non-agricultural income15. These results seem to contradict the intuition that as agriculture modernizes, forwards linkages should become more important, as countries develop an industrial food processing sector. This intuition, though, can be supported by looking at the relative size of the backward with respect to the size of agriculture. Additional regressions reported in Appendix I show that both the ratio of forward linkages to agricultural GDP and the ratio of backward linkages to agricultural GDP grow with development as measured by per capita income. These latter results altogether convey that although linkages fall with development, the size of these linkages fall less than the decline in the relative size of agriculture.

These results are rather surprising, and deserve a closer inspection. We need to compare the forward and backward linkages of agriculture, with the equivalently calculated linkages of other sectors we present the sectors with the highest forward and backward linkages for two countries: one a developing country with high backward linkages, Bangladesh; and the other a middle-income country with low linkages for agriculture, Chile. The table is very revealing, because it reinforces the results previously found. The backward linkages, the focus of early development economists, not only are higher at earlier stages of development, but these linkages are among the highest of any sector at earlier stages of development. In the case of Bangladesh actually the highest, since Rice Milling is totally linked to agriculture, and agriculture itself is second in the ranking. Furthermore, even in middle-income countries, agriculture still has high backward linkages, as in the case of Chile, where the sector represents 4% of national GDP, but ranks 10th out of 71 sectors in importance of its backward linkages.

The section above points to agriculture as a key sector to promote development as its backward linkages are highest at earlier stages of development. Therefore, a development strategy that promotes agriculture will indirectly "pull" the rest of the economy towards development. However, looking at output linkages may not be the key for the design of longterm development policies. As Timmer states, part of the controversy of the role of agriculture in development stems from the fact that structural transformation is a general equilibrium process that can not be explained by looking at agriculture alone.

López et al. explain the process of reduction of the relative size of agriculture, i.e. structural change within the context of a two sector endogenous growth model. The authors show that even when labor productivity grows at the same rate in agriculture as in the industrial sector, and even if there is no declining income demand elasticity for the agricultural good, agriculture will relatively contract because one of its production assets can not be indefinitely accumulated. This asset is of course a renewable resource, which can be understood as water, land, soil quality, etc., which is limited in its supply by nature. The income maximizing path is achieved by equating the returns to assets. The relative contraction of agriculture, and the migration of assets from agriculture into the industrial sector is a result of this equating of marginal returns to assets that can be employed in agriculture or industry, like labor.

The equation of marginal returns to assets, that at the aggregate level are always decreasing can be considered as the *golden rule of development*, not the promotion of the sector with high multiplier effect. While this has rather obvious policy implications and one that most economists would recommend intuitively, unfortunately if one looks at the "import substitution" era one finds the most egregious anti-agricultural bias and incredible disparities of returns to assets within the rural and urban world. For example, in China and India, the pursued industrialization policies implied a strong pro-urban bias leaving the returns to investment in the rural areas with much higher rates of return. In a very influential work, that is partially responsible for turning the anti-agricultural tide Krueger et al. with main results summarized in Krueger et al. show that on average over 16 developing countries for the period 1975-1984 the exchange rate overvaluation and import tariffs on industrial goods were the equivalent to at least an 11% export tax on each country's agricultural export.

Another important lesson that can be obtained from a general equilibrium development analysis, is an interpretation of poverty traps, i.e. those countries that are unable generate enough savings to lead them into the development path. These countries are characterized by overstretching their natural resources, a manifestation of this overstretching being an excess of labor resources employed in resource using sectors. The general equilibrium view suggests two ways to exit a poverty trap: On one hand, there is implicitly low productivity in the industrial sector, and investing in non-agricultural productivity can help increase the demand for labor in the industrial sector and reduce the pressure on the resource. However, another possibility is to increase agricultural productivity to make the operation sustainable. Today most of the rural poor live in fragile tropical ecosystems, and where labor is not a major economic constraint. Most of the human and financial resources, public and private, that are devoted to agricultural R&D, are employed in research that is relevant for capital intensive and temperate climate agriculture; i.e. which are the needs of developed countries' agriculture. The corollary is that although agriculture may play a key role in development, as we have argued above, this role may not be played if the agroecological conditions places limits to the expansion of the sector.

Agriculture's Pro-Poor Role

Not all growth experiences are equal. There is a growing focus on the importance of a "propoor growth" defined as growth an increase in average income purchasing power that is also accompanied by an improvement in the distribution of income. To help in the understanding of a formal definition of the concept, let us imagine a rural world with only 100 inhabitants. One person owns the only farm, and earns 101 monetary units, while the rest are 99 laborers that earn 1 currency unit each. The above example serves to highlight some important facts about growth. First, when the initial distribution is extremely unequal, growth is not enough to reduce poverty, even if there is "trickle down" of the benefits of growth, as in case (*i*). Given that many examples of growth that resemble cases (*i*) and (*ii*) have been documented, with limited or no reduction of poverty, is that researchers have turned their attention to "pro-poor growth". In our example assume case (*iii*) all incomes grow by 1. In case (*iii*), growth is also exactly 50%, but the distribution of income has improved, and more importantly poverty has been reduced to 0%.

There is ample theoretical support and empirical evidence that suggests that agriculture is pro-poor, and that growth based on the expansion of the sector is pro-poor growth. We can identify four main channels by which agricultural growth helps poverty alleviation:

- *Directly increasing the income/own consumption of small farmers*: Small holders are usually not only inadequately endowed with land, but usually also lack other assets, like physical and human capital, and thus are usually poor. Expansion of the agricultural sector may benefit also the small-holder sector and pull some of them out of poverty. Small farms are, with respect to capital and land utilization, labor intensive own family labor, and therefore are likely to benefit from technological progress that is labor intensive. When land distribution is equitable, it will be the case that expansion of agriculture will benefit the small-holder sector; when the land distribution is inequitable there could be agricultural growth fully based on large farm output expansion, in which case the small holder sector would not necessarily benefit.

- *Indirectly by reducing food prices*: Most measures of poverty are based, directly or indirectly in the cost of access to food. When the price of food is reduced, there is a two way accounting improvement in the welfare of the poor. In the first place their real income increases, and more so than the wealthier, because food is the main component of their consumption basket. At the same time, the poverty line which is usually used to measure poverty is decreased, which acts to alleviate poverty. In a completely open economy without any additional transaction costs, the price of food should not be affected by agricultural growth in the same country. However, not all food is tradable, many perishable vegetables are for all purposes non-tradables. In addition, the bulk of cereal staple foods produced and consumed by the poor in rural areas are traded in local markets which, due to high transactions costs are disconnected from larger including international markets.

- *Indirectly by increasing the income generated by the non-farm rural economy*: The rural non-farm economy in most regions is either: mostly composed of goods and services that directly serve agriculture, or indirectly depend of the demand of those tied to agriculture. The more disconnected the rural economy is from urban markets, the more dependent is the rural non-farm sector on the income generated by what is usually the main engine of the rural economy: agriculture. Hence, agricultural growth can increase the demand for the goods and service of the rural non-farm sector and help pull out of poverty households tied to this sector.

- *Indirectly by raising employment and wages of the unskilled*: Agriculture is usually intensive in unskilled labor. Thus, agricultural growth through an increase in unskilled labor demand will increase unskilled employment and or the wages of the unskilled, most of which are poor. There is here a general equilibrium effect, because raising the unskilled wages in agriculture pushes upward the unskilled wages in urban areas also. This latter general equilibrium effect will be higher the more integrated rural and urban labor markets are.

Given these channels through which agriculture reduces poverty, it should come as no surprise that the overwhelming empirical evidence shows that agricultural growth is not only pro-poor, but more pro-poor than other sectors of the economy. The key in this fundamental result lies in that all four channels described benefit the poorest households of the economy more than the rest. In a seminal study Ravallion and Datt used a long time series data from India, to explain poverty with output from the different sectors of the economy. They found a large elasticity of poverty with respect to primary output with the sector being more effective at poverty alleviation than industry. Kakwani used the additive property of the most popular poverty indicators, the FGT class, and decomposed the effect of sectoral growth in poverty. Using the information of a 1985 Côte d'Ivoire household survey she shows that the elasticity of poverty with respect to agricultural output is much larger (-1.8) than other sectors such as services (-0.1) and industry (-0.1).

Thorbecke and Jung used the additive property of the FGT poverty measures and a Social Accounting Matrix to decompose the contribution of each sector to poverty alleviation. The authors apply this methodology to Indonesia, and find that the primary sector has a larger contribution to poverty alleviation than the industrial sector, and slightly larger than the services sector. Khan applies the same methodology with data from South Africa and reports similar evidence, agriculture is the sector with the largest contribution to poverty alleviation.

The fact that agriculture is more pro-poor than other sectors seems to be substantiated in poor countries like India and Côte d'Ivoire; as well as in middle income countries like South Africa. It is reasonable to expect that at different stages of development different channels dominate the pro-poor role of agriculture. The four channels described above are ordered in their likely importance from earlier stages of development, a ranking that can be corroborated by the scattered evidence. In very poor countries, most agricultural output comes from small holders, and therefore, this is likely the most important pro-poor channel of agriculture, for example, for poor Sub-Saharan nations. This intuition is consistent with the results of de Janvry and Sadoulet who show that following a technological improvement, the direct effect of technical progress on poor farmers is the main poverty reducing factor in Sub Saharan Africa, larger than price and employment effects. In Asia employment effects dominate, while in Latin America indirect effects on the rest of the economy are more important in the poverty reducing effect of agricultural technological progress. The food price effect is also likely to be more important in poorer regions, because they are likely to have a food basket highly dominated by 1 staple crop, like in some poor Asian nations. Timmer shows that in the case of Indonesia 80% of the variability of the poverty to growth elasticity can be explained just by changes in the real price of rice.

Note that when the price of rice falls, it adversely affects poverty by reducing income of poor farmers; however, when the price fall is caused by gains in productivity this is not necessarily the case.

As rural economies develop, the non farm economy becomes more sophisticated and diversified, and as we show below more important within the rural economy. Therefore, this indirect effect becomes more important. On the other hand, with development, the rural non farm economy becomes more tied to the rest of the economy, diminishing this indirect effect.

Thus we also expect in very poor economically isolated rural economies, the non-farm sector to depend heavily on agricultural productivity, income and demand. This is consistent with the very high demand multiplier from agriculture found by Haggblade et al. in poor Sub-Saharan rural economies; but also consistent with the high indirect poverty elasticity of agriculture found by de Janvry and Sadoulet for Latin America in general. Finally, as agriculture makes its transformation into commercial farming, the direct effect of the sector on small and poor farmers becomes much smaller, but commercial farmers are employers of unskilled labor; and concomitantly with the commercial farm transformation and their interaction with world markets, generally rural economies become interlinked with the rest of the economy, in particular with urban markets. These are the preconditions for agriculture to have an important effect on the employment and wages of unskilled workers.

Anríquez and López show that in Chile, where the lion share of agricultural output comes from medium and large commercial farms, agriculture is still more pro-poor than other sectors of the economy, and that this poverty to agricultural growth elasticity is mostly explained by the effect of agriculture on unskilled labor markets.

Agriculture and Rural Development

To what extent is rural development the same as agricultural development? In other words in which case does a sectoral driver to rural growth is indispensable? To answer this question we need to have an idea of the importance of agriculture within the rural world. There is no direct measurement to uncover the importance of agriculture within the rural economy, but we will show alternative roads which indirectly point to that measure. Agriculture, is an important component of most rural economies especially in the developing countries. It was shown above that the size of agriculture within the local economy is sometimes used to define rurality. Therefore, any successful rural development strategy will contain an agricultural development component; but they are not the same thing.

While agricultural development aims at improving the welfare of populations through sustained improvements in the productivity of the agricultural sector, rural development aims at the improvement of welfare of rural populations through the sustained growth of the rural economy, which includes agriculture, but may not be its only component and not necessarily the most dynamic.

Ideally, if we wanted to know the importance of agriculture within a rural area, we would look at the GDP or output figure for that region, and measure the share of agricultural output within the total value added of the region. Unfortunately only few countries have aggregate output figures available by region state or province, and even at this regional level, output is aggregated for both rural and urban areas.

We show above that the importance of agriculture within the national economy falls with development, but this does not necessarily mean that the relative importance of agriculture in the rural economy also falls with development (given that rurality also falls with development). Poorer countries, with lower per capita income, and with higher incidence of poverty, not only are more rural, but in their economies agriculture has a higher relative weight. Thus, the poorer the economy, the more important agriculture is for its rural and overall development.

Another road to measure the value of agriculture in the rural economy is to examine the share of income from agriculture to total rural income. This can be done, thanks to household surveys that measure income, most of them implemented in developing countries since the 1950's to understand poverty. Unfortunately, the way researchers have measured rural agricultural income *vis-à-vis* non-agricultural income also known as rural non-farm income has varied too much. Examples of these conceptual inconsistencies are: to add remittances which in countries like Pakistan and in Central America can amount to more than 5% and even 10% of household incomes to rural non-farm income, when remittances are not *rural* sometimes not even national income. Other inconsistencies arise with income from wages of agricultural labor. Some authors add wages accrued in farms outside their own to non-farm income, when it is clear that this is agricultural income. Further problems arise when wage income due to the absence of information can not be assigned to any particular sector, in these cases all of wage income can be added to either the farm or non-farm sector.

Some authors even add food sales to non farm income. Then there are gray areas, like the way in which to value own agricultural consumption; or income from fishing and forestry that could be added to either farm or non-farm income. These conceptual differences and inconsistencies highlight the need for comparable measures of agricultural and nonagricultural rural income to make further meaningful cross-country comparisons and studies.

However, for our purpose of uncovering any links between rural agricultural income and development, these measures, alas inconsistent, will be helpful. For this study we have collected share of rural agricultural income for 120 country-year combinations. Three country-year combinations are repeated with different figures, and we keep all the information as we are in no condition to discriminate the inaccurate information. The main sources are, FAO and Lanjouw and Feder, that made comparable recollections of rural nonfarm income shares, and our own collection from more than 30 monographs, mostly from rural household survey studies. This database contains information from four continents and it spans the period 1950 to 2002. Table 6 summarizes the average share of agricultural income in total rural income by continent.

This dataset by no mean should be considered a representative sample, because this sample except for Japan, contains only developing nations. Furthermore, some countries have better information and are over-represented; this is the case for example of India and Pakistan in Asia, and Mexico in Latin America. However should give us some idea of differences by continent. Perhaps a more meaningful sample is obtained by including only the more recent observations, from 1990 onwards, where we can see that both in Africa and in Asia, agriculture accounts for more the 3/5 of total rural income, while in Latin America, agriculture accounts for 56% of rural income explores the existence of an empirical regularity between level of development and the relative size of agriculture in the rural economy. The Table shows that as per capita GDP increases, the share of agricultural income in total rural income tends to fall, although at middle income level medium for the sample the share seems to lie relatively stable. Per capita GDP may be a good indicator of overall development for a country, but not necessarily an equally good indicator of rural development. We propose as a better indicator of rural development an alternative measure of income: agricultural GDP per rural inhabitant. Obviously both income measures are highly correlated correlation coefficient of 0.93 in our sample, but the proposed income indicator has the correct denominator and its numerator is very likely highly correlated with the ideal one. The lower panel of the table shows that when the share of agricultural income in total rural income is ordered by agricultural GDP per rural inhabitant categories, there is clearly a diminishing trend for the share as income increases.

The negative relationship between rural development and the share of agricultural income can also be confirmed by examining time series data for particular countries. In Japan for example the share of agricultural income in rural income dropped from 82% in 1950 to 30% in 1980; similar drop can be observed in India, 92% to 62% in the period 1958-1994; and in China from 90% to 67% in the period 1980-1997. In Bangladesh the share of agricultural income in rural income actually went up in the period 1963-1982 from 82% to 92%, but real income also dropped in this Asian country during this same period agricultural GDP per rural person fell from 103 to 91 US$.

Thus, the cross country as well as the time series data indicates that a feature of *rural development* is that the share of agriculture in rural income falls with development. Nonetheless, as indicates, agriculture is the main source of income for rural households, responsible for more than three fifths of total rural incomes.

Spatial Constraints to the Promotion of the Non-Farm Economy

As we have established that a feature of rural development is a relative increase of the nonfarm economy, it is important to characterize the sector. In general, in the rural space we find activities that require proximity to the point of extraction or production: primary activities like agriculture, fishing, forestry and mining; industry that may benefit from proximity to the sources of raw material like food including fisheries products and other processing fuels, timber, and metals; and services that need proximity to the resource like eco and agro tourism. In less developed countries scant evidence indicates that most non-farm activities i.e. non primary are closely linked to agriculture and cluster regionally in small and medium-sized towns, Haggblade et. al. Davis et al. find similar evidence, as they show that proximity to infrastructure is robustly correlated with higher non-farm income in a cross section of developing countries.

The realization that non-farm activities enjoy economies of agglomeration and tends to cluster throughout the rural space has motivated the promotion of rural development strategies that promote the non-farm sector but with regional focuses. One early such strategy was the clusters promotion, which focused on the development of regions specialized in one industry or non-farm sector i.e. wine industry or steel industry cluster. More recently, the territorial approach to rural development has proposed a more integral approach by seeking the promotion of both agricultural and non-agricultural activities jointly at a regional level, de Janvry and Sadoulet. Both strategies presume the existence of rural-scale agglomerations that make viable the promotion of a particular non-farm sector.

A successful territorial approach strategy requires the existence of the agglomerations and/or infrastructure which will facilitate the reaping of benefits from economies of agglomerations and scale we present the spatial distribution of different countries in Sub-Saharan Africa, using the previously described GRUMP-CIESIN database. In panel A we see the spatial distribution of population in South Africa. In this country there is a smooth gradient of population densities, and we thus have that people leave in areas of different densities, from large cities like Johannesburg and Cape Town to sparsely populated rural areas. In the rural space we have the presence of large medium and small towns. Therefore in South Africa development projects that anchor in large rural towns have the potential to transmit to smaller town and to less populated areas.

Conclusions

The last few years have seen an increased attention of the international development community on agriculture and rural development. If the data for the last 3 years constitute a real trend then we are witnessing a real increase in official development assistance to agriculture and rural development. The emergence of private donors and foundations such as the Gates foundation with keen interest in transferring resources to agriculture is a welcome development. Major multilateral donors are looking at agriculture as an engine for poverty reduction for most developing countries and regions and a fundamental component of a growth and poverty reduction strategy for the poorest, agriculture based economies. There seems to be increased interest by domestic and foreign private capital including multinational agro-industrial firms for investments in sectors up and downstream of production agriculture. The potential of agriculture as a source of bioenergy promises to attract further attention to the potential of the sector to produce and address global food and energy needs.

The present paper provides analytical evidence and arguments in support of the proposition that agriculture and the rural economy are fundamental for obtaining substantive and sustainable gains in the fight against poverty. Even using the existing, highly imperfect measures of "rurality" which very likely underestimate the economic importance and dimensions of the rural space and the extent of rural poverty poverty is intrinsically linked to rurality. What the evidence shows is that even when population movements and demographics reduce the share of the rural in total developing country population, poverty will still be a predominantly rural phenomenon. The experience of Latin America the most urbanized of the developing region is instructive in this regard: even in countries where the number of rural poor are less than the number of urban poor, the poverty rate is nevertheless higher in the rural areas.

The paper also shows that the analytical underpinnings of a development strategy with a strong anti-agricultural bias do not hold in light of new data and evidence. Agriculture has strong links with other sectors in many countries and, a productivity-induced agricultural expansion can "pull" other sectors with it, increase economic activity and employment opportunities in the rural areas. The paper also shows that, while a regional or spatial approach to the development of the rural economy is a reasonable proposition for countries with good infrastructure, functioning labor and other markets, there is substantial scope for a sectoral entry point for rural and possibly overall development and poverty reduction. This proposition is further corroborated by the results of Davis et al. who show that despite income diversification by rural households; households in the lowest expenditure categories have a larger share of agricultural in their total income than households in higher income groups.

The balance of entry points for rural development and poverty reduction is bound to change, as food systems change both globally and, especially in developing countries. However, such a change hides both opportunities and threats. First, the role of agriculture in the structure of the economy will decline in the process of development. However, the data on the agricultural transformation shows that the labor share of agriculture declines much slower than the share of agriculture in national GDP. Unless policies and investments are put in place to foster agricultural productivity, there is a danger that the decline of agriculture will be accompanied by increased rural poverty some of which will find its way into the urban areas. At the same time, policies and programs which increase the human capital of the rural poor and allow them to enter a more remunerative labor markets are powerful tools to ensure a smooth transition of people out of agriculture without increasing poverty. At the same time a transformation of the agricultural sector itself is underway. Growing per capita incomes have increased and will increase further the demand for high value food products and for the quality and safety characteristics of foods. Such shifts in demand are reflected in the structure of the food systems even in some of the poorest developing countries towards the spread of the modern food chains and consolidation of the production, distribution and retail segments of food markets. The role of the more traditional "chain" will shrink overtime. Therefore, while increasing productivity for food staples oriented agriculture will still be an important anti-poverty entry point for many poor areas in the world, the high requirements in terms of skills and capital by the "new agriculture" point to the need for increased access to modern technologies research and extension systems which promote the higher quality and safety standards demanded by consumers.

Will trends towards domestic market integration, and globalization affect the strength of the links between primary agriculture and up-stream and down-stream sectors? On one hand, market integration will sever the links between agriculture and the local economy, but will open new links to larger markets regional, national or even global. Opening the agricultural sector to foreign but also domestic competition will challenge the position of smallholders in national markets. The non-farm rural economy will also be subjected to competition pressures from cheaper consumer goods distributed by supermarkets and similar retail outlets as they expand to the rural areas of developing countries. Small scale and artisanal food processing may give way to more organized, capital intensive processing plants and a similar scenario may be thought for other activities up or down-stream primary agriculture such as transport or input processing, etc.

Chapter 4

Agricultural Production, Agricultural Land and Rural Poverty

Rural areas dependent on agricultural income, are often among the poorest in developing countries. However, little distinction is generally made within the agricultural sector. This lack of distinction hinders targeting of agricultural investments towards poorer farmers. This paper illustrates, using a production function analysis with flexible marginal returns, how agricultural production activities and returns to agricultural production factors differ by poverty level in the case of Madagascar. The results show that access to primary education is relatively more beneficial for poorer agricultural households while additional secondary education has no effect on agricultural productivity. Returns to agricultural inputs are significantly higher for poorer agricultural households. Land inequality increases as land sales markets benefit the richer households and as the rich engage more in extensification while rental markets improve agricultural efficiency and may thus benefit poor and rich alike. Land titling has little effect on improved agricultural productivity. More formal land titling is therefore not sufficient to change the bad performance of agriculture of the last decades.

Introduction

The objective of this analysis is to study the link between agricultural production and rural poverty and to quantify the determinants of agricultural productivity in Madagascar. This topic is of interest for several reasons. First, the majority of the population in Madagascar are rural households for which a large part of their incomes originates from agricultural production. Agricultural performance and growth is then crucial in any poverty reduction strategy. Second, a well functioning agricultural sector has multiplier effects on non-farm activities and might therefore contribute to growth in other sectors and thus in the whole economy. Third, economic reforms that started in the late 80's, and which are still going on, did not induce a substantial increase in agricultural production. It is then important to better study the constraints that still impede high crop productivity and, therefore, rural income growth. As such, the study hopes to be an analytical background paper for the rural development strategy that is currently being developed under the PRSP.

This study aims to be a quantitative economic study. One of its contributions is the explicit distinction within the agricultural sector of poor and non-poor farmers. Unfortunately, good recent nationally representative data on agriculture are lacking in Madagascar. The national agricultural census dates back to 1984, a period when agricultural reforms were still under way. None of the recent national household surveys have good agricultural data. The most complete agricultural information is available in the comprehensive EPM of 1993. Therefore, we will base ourselves on the agricultural households of that dataset in this study. As there have been no major structural changes in agriculture since 1993, we believe that this dataset still provides a good overview of the current agricultural situation in Madagascar. If necessary, we refer to more recent datasets and studies, but not representative at the national level, in the descriptive analysis.

The structure of this study is as follows. We will start with a brief review of the economic reforms in Madagascar since independence. This will be followed by a descriptive analysis on the relationship between poverty and agriculture. The next section will present the characteristics of agro-ecological regions and their relation to poverty. Then the main analytical results on the quantitative importance of the different determinants of agricultural production are presented. The study finishes with conclusions and policy implications.

Overview of Economic Reforms in Agriculture

In this section, an historical overview of agricultural policies in Madagascar since independence is given. This overview will help us to better understand the current agricultural policy environment. Agricultural market reforms in Madagascar resemble those in many other African countries which have gone through cycles of government interventionism and retreat. After Madagascar obtained independence from France, governments initially increased the intervention of the state in agricultural markets such that by the end of the 1970s, most trade in agricultural products and inputs was in the hands of the state. A reversal of policy took place in the 1980s with a transition from a state food marketing and distribution system to a liberalized market. This transition, however, was very gradual.

During the First Republic small traders together with the parastatal Bureau de Commercialisation et du Stabilization du Riz, organized the marketing of rice. The BCSR fixed minimum and maximum prices, provided credit to farmers, and organized rural associations. During this era, agricultural policies were focused on increasing the rice area through large irrigation infrastructure schemes in areas such as Lac Alaotra, Marovoay, and the Delta of Mangoky. This was combined with agricultural extension efforts focused on the use of modern inputs and improved equipment for rice cultivation. Consequently, from independence to 1968, rice production increased by almost 50%. This change was the combination of increases in area as well as productivity. It seems that appropriate macro-economic policies, favorable terms of trade, and active participation of private operators largely contributed to the good performance of the agricultural sector during this period.

In 1972, the new socialistic government wanted to get rid of the private marketing sector that was perceived to be predatory. Thus, the government created a national monopolistic parastatal - the "Société d'Intérêt National des Produits Agricoles" that was responsible for assembly, transformation, and marketing of agricultural products. A tax system as well as economic barriers were put in place in order to allow each Fokontany to benefit from agricultural production and to control product movements. The taxation systems penalized especially export crops. For example, from 1975 to 1983, coffee producers only received 40% of the world price and those of vanilla and cloves 25%. The consequences were a net reduction of export revenues. On the other hand, domestic food prices were subsidized and artificially kept low which led to low domestic production and a surge in rice imports.

As the government was unable to continue to pay for food subsidies, liberalization began in 1983 when the state officially abandoned its monopoly on the commerce of agricultural products. The initial liberalization measures implied that agricultural trading was open to everybody except in the two main production areas the plains of Marovoay and Lac Aloatra where two government agencies, FIFABE and SOMALAC, could continue their monopoly. The roles of these two state companies were redesigned only in 1989. At the beginning of the reforms, floor and ceiling prices were maintained. In June 1985 a government decree fixed the floor price of paddy rice, but completely removed the ceiling price. In reality, the government still effectively controlled domestic rice trade until 1986.

In the 1990s, the government became more concerned with the effect of price policies on producers. For example, in 1991 the government introduced an import tax of 30 per cent on rice to protect local production. However, the government has frequently changed its taxing policy on rice. In 1996, the tariff on rice was 30%, going down to 10% in 1997 and to 5% in 1999, before increasing to 30% in 2000. On the other hand, the government occasionally granted tax exoneration for certain companies and shipments to assure a steady food supply. The exchange rate was made flexible in 1994 and let to a further increase in producer prices. Other liberal policies were put into place. The import licensing system was abolished, export taxes for cash crops were gradually eliminated, and the state disengaged of the maintenance of major irrigation areas, transferring responsibility to producer organizations. The current situation in agricultural markets can be described as one in which private traders have been given free rein to set prices and move agricultural products around the country and in which there is little state intervention.

On the Links between Poverty and Agricultural Production

While there have been some tangible macro-economic improvements in the last decade, around 70% of the population in Madagascar is currently still considered poor. Moreover, poor households live overwhelmingly in rural areas and depend mainly on agriculture for their livelihood. Hence, it is clear that the dismal performance of agriculture in the last decades is strongly related to poverty in rural areas.

In this section, we look in detail at these linkages between poverty and agricultural production. We use a money measure of welfare and poverty, i.e. aggregate household expenditures per capita, as developed by Razafindravonona et al. The use of expenditures as a measure for poverty has been shown to have considerable advantages: they are likely to fluctuate less than income over time and they are often more truthfully reported than income. To be able to better compare expenditures across households in the country as a whole, real expenditures per capita are used. Means of agricultural production and agricultural production factors are calculated by quintile of total expenditures, corresponding to five classes of poverty, starting with the first quintile, the group of the poorest, to the fifth quintile, the group of the wealthiest. The discussion is organized around the three main agricultural production factors: land, labor, and capital. The data used for this analysis come out of the "Enquête Permanente auprès des Ménages" done in 1993. This was a nation-wide comprehensive survey with information, among others, on household characteristics, consumption, health, education, and time occupation as well as on agricultural production. While the dataset and the sampling frame is not designed to produce accurate statistics on specific agricultural crops in Madagascar3, we are confident that the agricultural data have enough observations and variability to allow us to give an approximate overview of the situation and to establish causal relationships at the national level. The EPM sample consists of 4508 households. As the focus of our study is agricultural production, all households without land in cultivation are dropped. This reduced the sample to 2953 households.

Malagasy households earn on average 1.5 million FMG per year out of agricultural production, i.e. around $230 per household5. While the value increases by quintile, it does not increase monotonically over quintiles. However, this seems to be driven by some outliers in the fourth quintile as the median does show a monitonical increase. The richest quintile shows, on average, agricultural income that is three times as high as for the poorest quintile. If we calculate this income per unit of land, the difference is less but the rich still have higher income per unit of land than the poor do. Hence, the poor produce less agricultural products in total and their land is relatively less productive. We will discuss reasons for this further on.

The type of crop that is grown seems to matter to welfare. Richer households are more likely to grow cash crops, such as coffee, cloves and vanilla, than poorer households are. These households are dominant in the East and the North East regions. Households that grow vegetables are mostly located in the Highlands, and in the geographical area surrounding large urban consumption centers. 9% of the sample was involved in this type of activity6. Income from fruits was reported by 26% of the households. In both cases, households that grow these higher value crops are found relatively more in the richest quintile.

Agricultural land ownership turns out in most rural poverty regressions to be a significant determinant of welfare. This means that land access can potentially make an important contribution to the welfare of the poor. Hence, we will look in detail with the data at hand at this variable. We will discuss consecutively the link between poverty and land ownership, agricultural land expansion, titling, and land markets.

It seems that lack to land is highly correlated with lack of other productive assets including education which imply that the agricultural landless are the poorest group in the population. Unfortunately, there is little information on the landless in the EPM dataset so we limit ourselves to discussion of the households that own land. The average area owned per household is slightly higher than 1 ha, relatively small compared to other African countries. We notice a range from 84 are per capita for the poorest quintile compared to 181 are for the richest quintile, almost three times as much. One third, on average, of this land is the more valuable irrigated lowland. The land distribution is more unequal if we take into consideration this quality difference: almost 40% of the land that the richest quintile owns is irrigated compared to only 27% for the poorest quintile.

While richer households own absolutely more upland than the poor do, upland crops are *relatively* more important for the poor. Poorer households then depend relatively more on crop output from tanety for their income as well as for their consumption. One interesting example is cassava. This crop is mainly grown and eaten by the poor. Ravelosoa et al. and Minten et al. illustrate the negative income elasticity for cassava and its characteristics as an inferior crop mainly consumed by the poor. Based on the same data that are used in this study, Ravelosoa and al. show that the income elasticity for cassava is around –0.88, meaning that if income goes up by 1%, cassava consumption declines by 0.88%. This makes it an ideal food crop to be used for targeted interventions for the poor.

Land ownership might not be the only indicator of agricultural prosperity in rural Madagascar. For example,

Freudenberg shows that ownership of cattle in Betsileo villages in Fianarantsoa was considered more important for wealth ranking than access to land. The same holds in most of Southern Madagascar. However, nationally, land ownership is a sign of wealth and is a significant determinant of welfare. By comparing agricultural land data from the national census of 1983 and the EPM of 1993, Dorosh et al. show that landownership might have become more unequal over time. Two reasons seem to explain this: existing land might be sold by the poor to the rich and agricultural expansion is mostly done by relatively richer households.

Freudenberger provides a good discussion on the link of poverty and land expansion in the forest corridor of the Fianarantsoa area. She shows that it is mainly the relatively richer families that expand land holdings by moving into the forest and clearing new parcels as they try to assure future land for their offspring. The poorer population has shorter time horizons and is unable to make this type of investment.

Agricultural land expansion

Due to high population growth in rural areas, demand for new agricultural land is high. Data from a survey by IFPRI FOFIFA in three regions in Madagascar show that while rice land area grew by about 5 per cent as an average for all regions over a ten year period, the area of cultivated upland increased by about 24 per cent.

Considering that the population in the survey regions grew by about 3.5 per cent per year over this time, or by more than 35 per cent over the previous ten years, the rate of growth in arable area is actually below that of the growth in population. The expansion of the agricultural frontier came at the expense of forest-, bush-, and grassland. These land types are usually held as common property resources but become, under Malagasy law, private property when cultivated by successive generations of the same family. On average for all regions, the largest losses were experienced for area under primary forests. Indicating the level of primary forests as 100 for 1987, the index fell to 67 in 1997. The loss in secondary forest amounts to about 28 per cent, whereas the loss in grass- and bushland was about 20 per cent. These numbers coincide with alternative estimations based on aerial photographs.

That the agricultural frontier can still be expanded in many Malagasy villages is evident from the survey data: 59 per cent of villages report that there is additional land available for expanding upland cultivation; 50 per cent of villages for expanding irrigated land, and 35 per cent of villages for both types of land. Despite this, the average holding of upland per household has declined over time. Clearly, growth in population has outpaced growth of agricultural land in many communities and pressures for agricultural intensification have tended to increase over time.

Land titling

Traditional and modern land rights co-exist in Madagascar. Legally, all non-titled lands belong to the State while in practice, land even in inhabited areas are often already allocated through traditional rights. It is possible to move from traditional to modern legal rights but the process is slow, cumbersome, and non-transparent. Under traditional tenure, land property may be individual or collective. However, traditional land tenure systems in most parts of the country mean individual property rights. In some areas, but not all, some socio-economic groups are excluded from land ownership. Evidence exists of the existence of a cast-based system where the lowest caste, often descendents of slaves, are socially isolated and effectively excluded from land ownership. Their only way to have access to land is through sharecropping or land rental agreements.

The modern system of land titling has generated its own conflicts and it seems that the titling process benefits the poor more than it benefits the rich. While the absolute number seem to be overestimated and might not always reflect legal titles in the name of the owner, the percentage of titled land is as high as 32% for the richest quintile compared to only 20% for the poorest. Hence, while the poor have access to less land, the land they own seem to have less secure legal rights as well.

Land markets

The EPM data show that while land sales markets are quite active as 3% and 10% of the households sold and bought land over the last five years respectively, most of land that households own, was inherited: 67% of the land was acquired through inheritance compared to 16% through purchases. Land sales markets seem to contribute to more unequitable land distribution shows, that during the five year period prior to the survey, 5% of the poorest quintile reported land sales compared to 2% of the richest quintile and 13% of the richest households bought land during that period compared to only 7% of the poorest quintile. 12 The same pattern is seen when one looks at the way land was acquired. Inherited land represents 62% of total land for the richest quintile compared to 76% for the poorest. While the poor start off with less land, they also participate less in land sales markets and they get less land free from the village, leading to an even bigger landownership gap between rich and poor.

While land rental markets seem to be more active in more densely populated areas, the IFPRI FOFIFA community dataset also seems to indicate that land rental markets seem to become more active over time. Based on this dataset, it was estimated that the lowland area under rental agreements increased over a ten-year period by 4%, from 7% in 1987 to 11% in 1997. Overall, upland areas are less rented out than lowland, but this practice seems to become more in vogue: rented upland area increased over the same period from 3% to 5%.

Labour

Agricultural production in Madagascar is very labor intensive, with the exception of cases where animal traction or tractors are used in general, still rather rare and of regions where labor extensive techniques such as slash-and-burn agriculture and direct seeding are still dominant. The importance of labor use for agricultural production has been shown by various studies on labor allocation and agricultural productivity. Rice with the labor intensive transplantation technique and the manual harvesting method is one of the most labor-intensive crops in Madagascar. For example, the average number of labor needed for one hectare of rice varies from 80 person-days for the Lac Alaotra zone to 400 person-days or more in the southern Highlands. The small average rice area by farm does not seem to justify individual investments in mechanaical riceplanters or harvesters14 while rental markets for this type of equipment seem to be thin or non existent. Other reasons for high dependency on human labor might be its comparative advantage with respect to modern imported equipments, as there seem often to be difficulties to find spare parts and as maintenance costs in humid areas might be prohibitively high.

Unfortunately, the EPM dataset contains little data on agricultural labor use or on the functioning of agricultural labor markets. However, it appears from other surveys that agricultural labor markets are quite active in Madagascar. Lapenu and Zeller mention that more than 50% of rural households in the IFPRI FOFIFA survey earned income out of wage income at some point during the year.

Female-headed households seem a little bit more represented in the poorer compared to the richest quintile. However, the difference is small. On the other hand, there do not seem to be major age differences between richer and poorer heads of households, implying that life cycle phenomena might not be important to explain welfare differences. There is no clear distinction between the poor and the rich in the number of households that report agriculture as their main profession ceteris paribus, i.e. given that they cultivate land.

Capital

Not surprisingly, poor have less access to and use less capital in agricultural production activities. They rely less on oxen or agricultural equipment to improve agricultural labor productivity and they rely less on input use mineral and organic fertilizer, pesticides, and purchased seed to improve land productivity. The major reasons seem to be seasonal or permanent liquidity constraints and less riskbearing capacity and thus less willingness to invest in inputs.

The EPM show that agriculture expenditures per unit of land are low, one of the lowest in Africa. Expenditures are around 161 Fmg per are, corresponding to an application rate of less than 8 kg of NPK or Urea per hectare17'18. Organic fertilizer use shows high variability between agro-ecological regions19. With the exception of some industrial crops use pesticides. In most cases, it seems that pesticides are only used when there is already advanced pest damage. The use of improved seed is also extremely low.

Overall, shows that expenditures on modern inputs are the lowest for the poorest quintile. Because utilization rates are so low and the average is driven by few observations, big variability for the four individual inputs across quintiles shows up. Aggregated over expenditures for inputs, the richer quintile applies 20 times as much input than the poorest quintile. The total value for agricultural equipment shows the same trend: it is seven times as high for the richest compared to the poorest quintile. When we control for the area cultivated the differences are less striking. However, the rich use, per unit of land four times and twice as much inputs and agricultural equipment respectively. Surprisingly, there is no difference in the number of draught oxen per unit of land and the difference in the value of agricultural "equipment" seems to come solely from mechanical equipment.

Farmer's access to credit seems an important indirect way to improve agricultural productivity in Madagascar. It has been shown in Madagascar and elsewhere that an program "Programme National de Vulgarisation Agricole" was the production of compost at the farm level increase in cash availability either through credit, non-farm income, or from off-season crops greatly contributes to increased input use and improved technology adoption. Major efforts have been done in Madagascar in recent years to improve formal credit access through the implementation of various mutual credit systems. As our data date from 1993, we can not see the effect of this. However, it can be stated with reasonable confidence that most of the smallholders in rural areas still remain outside of this formal credit system. The EPM data show that only 29% of households have used credit. The poor rely relatively a little more on credit than the rich (30% of the households compared to 26% for the rich). However, only 1% of the poor gets credit through the formal system compared to 6% for the richest quintile. Therefore, poorer farmers rely relatively a lot more on informal credit systems. There is evidence that interest rates are significantly higher in the informal sector. It is interesting to note that there are no major differences across quintiles for use of the credit, except for the highest quintile where a significant part uses a large part of the credit for nonagricultural purposes.

Surprisingly, the poor live less in areas that are prone to weather problems. 19% and 31% of the poor and the rich respectively lived in a village that was hit by a cyclone in the year previous to the survey. The areas in Madagascar where most of the cyclones strike are on the East Coast which are also the areas where much of the high value export crops are grown. Due to their close distance to the big ports, these areas have an easier access to export markets. The poor also live in villages that had less problems with drought: 29% compared to 40% for the rich.

The descriptive statistics by expenditure quintiles show that the poor in rural areas have less land and assets, have less access to infrastructure and institutions, and rely relatively more on family labor for their agricultural activities. Even if they are disadvantaged in almost all areas, the return to some of the production factors in agriculture might be higher for the poor. The essence of a strategy for reducing rural poverty is to set priorities in possible interventions. The analysis that follows is intended to shed light on the effect of some of these interventions on agricultural production. This will be done using production function analysis. However, not only economic factors and choice variables but agro-ecological conditions as well influence agricultural productivity and poverty. To get at some of these links for Madagascar, we will discuss in detail the regional characteristics and how they might have an effect on poverty and agricultural production. To this we now turn.

Poverty and Agro-ecological Regions

There is ample empirical evidence on significant regional poverty differences in Madagascar. As rural income is strongly linked with agriculture and agricultural production depends on the natural environment, poverty is strongly linked to regional characteristics. We will move away from the usual administrative delimitation, i.e. the Province, for this study on agricultural production and will use agro-ecological boundaries. The agro-ecological delimitation used in this study is based on soil characteristics, rainfall, topography, climatic data, and cultural and anthropological considerations. Graph 1 shows the proposed delimitation. We base the definition of agro-ecological region mainly on the results of three documents: the last agricultural census report and regional studies in 1993 and 1995.

North & North East

The North & North East region is made up by the province of Antsiranana as well as two Fivondronana from the province of Toamasina, i.e. Maroantsetra and Mananara. We distinguish two sub-regions: the Eastern and the Western part. Rainfall is generally abundant but sub-regional variability is noticed. The Western part is less humid than the Eastern part. In the latter, average annual rainfall may reach up to 3,500 mm as for example in the Maroantsetra area.

The Eastern part encloses two large areas of preserved forests: the Mananara Biosphere Reserve and the Masoala National Park. Soils are generally of poor quality and are not very well suited for agricultural production. This region includes also "Sambava, Vohemar, Antalaha" or the SAVA zone, where most vanilla 90% of national production and cloves 50% of national production are grown. 64% of the households that were interviewed reported having income from export crops. This region is also characterized by a high number of cattle in the sub-region of Antsiranana and in the Fivondronana of Ambilobe. Rice dominates agricultural production in the Andapa plain. This plain provides 85% of total rice production in the SAVA sub-region. The Western part is characterized by fertile volcanic and alluvial soils in the Sambirano zone and the area surrounding the Montagne d'Ambre; and less good soil in the Ankarana zone. Industrial sugarcane plantation essential oil plants, and cacao plantation are also found in this area.

East Coast

The East Coast is divided into three sub-regions: the Northern, the Center, and the South. It consists mainly of the coastal area of the province of Toamasina and Fianarantsoa, and the Fivondronana of Tolagnaro in the province of Toliary. Rainfall is abundant during 200 days over the year with an annual average ranging from 2,800 mm to 3,500 mm from the South to the North. Temperature is no constraint to agricultural production. However, the effect of cyclones, and inherent flooding, limits lowland cultivation and intensification.

The region is specialized in export crops such as coffee, cloves, pepper, and different exotic fruits such as litchis and bananas. Rice cultivation is mainly done on the hillslopes using slash and burn methods. It is estimated that 48% of total cultivated area in this region is occupied by rice.

South & South West

The region is divided into two sub-regions: the South and the South West. Annual raintall only averages 400 mm in the South and 600 mm in the South West. This is clearly the major constraint to agricultural production in the region. This is also the region where the dominant agricultural production system is based on tubers instead of rice. In some areas in this region, the farming system is based on extensive livestock production. During the last years, many NGOs and projects started rural development activities as this region is perceived as one of the poorest and most food insecure areas in Madagascar.

With the exception of few areas such as Amboasary and those with irrigated system, soils and climate are barely suitable for agricultural production. With a production of 6 kg of rice equivalent per are, the region has the lowest total value product compared to other agro-ecological regions. Soils are sedimentary, rich in calcaire, or sandy in the semi-desert zone. Cassava, sweet potatoes, and maize are the most important food crops. Rice occupies only 32% of total cultivated area. Sisal production is the specialization in the irrigated Mandrare plain.

Beside the dry conditions, various other factors might be associated with poverty in this region. First is the low education level of the household head as only 34% of them finished primary school. Rural insecurity might also hinder development and poverty alleviation. Table 6 shows that 70% of the villages in the sample reported cattle theft during the previous year. 95% of the sample still report to make their livelihood from agricultural production. This is the highest percentage with the Littoral West and the Mid West region.

However, livelihood in agriculture includes those farmers who are specialized in livestock production, which seems to be the case for a significant number of farmers in this region.

Littoral West

23 The instability of the coffee price during the last five years illustrates this point very well. From 1997 to 2001, the farmgate price of coffee dropped by 700%, plummeting from 7,000 Fmg to 1,000 Fmg per kilo, mainly caused by the change in the world price.

This region includes the Fivondronana between Antsalova in the South to Ankaizina in the North. The geography allows for many large alluvial plains, suitable for lowland cultivation. Average rainfall seems to be sufficient, ranging between 1000 to 1500 mm from the North to the South. The major agricultural constraint originates from the long dry season: from April to October. Temperature is relatively high, with an average of over the year. Mangrove forests dominate the vegetation in the littoral and in the large bays of Bombetoka and Mahajamba. The ferruginous and vertisol soils in the plateau zone hinder high agricultural production. The baibo, alluvial soils along the rivers, are the most fertile land in this region.

Rice is the most important crop and occupies 75% of the total cultivated area. This region includes the large plain of Marovoay, which is the second largest irrigated plain in the country. Its presence results in a relatively high percentage of irrigated land in the region, with 47% even just higher than in the Highlands. In Marovoay and Ambato-Boeni, *jeby* rice, harvested in November i.e. during the lean period for the rest of the country is the dominant crop. *Asara* rice, harvested in May, is important in the mid altitude plains. Slash and burn practices are frequent in the Eastern part of the region.

A low 3% and 9% of the farmers produce export and vegetable crops respectively. The cash crops in this region include mostly industrial crops such as cotton, sugarcane, and tobacco. Some Fivondronana in the Northern zone are specialized in tomato and onion production and provide most of the supply for the capital. Constraints to improved agricultural productivity and higher incomes might be the lack of infrastructure such as road and market facilities as well as low levels of education.

Highlands

This is a large and diversified region. The Highlands are the most densely populated region of Madagascar, with a population density of around 150 per square kilometer. This region is well off with respect to access to institutions. It is characterized by a relatively high level of education, as 81% of the heads of households finished primary school. They are also well served by agricultural extension: 45% of the households lived in a village where this service was available in 1993. Four sub-regions are distinguished within the Highlands: North, Center, South and the Lac Alaotra region.

The Northern subregion includes all the Fivondronana close to the Tsaratanana Mountains. This is a relatively remote area despite its underexploited agricultural potential, such as the Bealanana plains. The Center and the Southern sub-regions, starting from the Fivondronana of Ankazobe to Fianarantsoa is the most well developed region in terms of infrastructure and the presence of institutions. It is also characterized by the existence of three large urban consumption centers, i.e. Antananarivo, Antsirabe, and Fianarantsoa. The Center has a higher average rainfall compared to the South.

Eroded and hilly ferralitic soils dominate the plateau while hydromorphic soils are frequent in the lowlands. The farmers in this area have a long agricultural tradition in rice cultivation. Rice occupies almost all of the lowland area. On the upland, farmers grow a large variety of crops such as maize, beans, and potatoes in the Center and cassava and sweet potatoes in the South. This is also the area where temperate crops such as wheat, barley, and temperate fruits are grown. However, upland crops do not show a high productivity except for the volcanic and alluvial soils in the Antsirabe region.

The Lac Alaotra sub-region has an altitude of around 800 – 1200 meters above sea level. It is characterized by the existence of a large plain with modern irrigation infrastructure. Three Fivondronana make up the sub-region: Ambatondrazaka, Amparafaravola, and Andilamena. This is a rice surplus area and supplies annually around 200,000 tons of paddy-rice to the urban areas of Antananarivo and Toamasina. The area has relatively good intra-zone road infrastructure. There are also many agricultural institutions present such as agricultural credit systems, agricultural research stations, extension, etc. and different administrative services such as a land property registry and courts. The plain is characterized by hydromorphic soils, clay textured, and is very well suitable for rice production. However, initially rich in organic matter, the soil currently presents signs of nutrient deficiency. The plateau around the vast plain has ferralitic soils while the mountains are characterized by very poor and fragile soils, subject to spectacular erosion sites or "lavaka", resulting in serious sedimentation in the lowland. One of the agricultural issues in this area is to solve the externality problem of halting hillside erosion to protect lowland rice areas.

The Highland region shows the highest TVP of 2.2 million Fmg per hectare. Land scarcity it has, with a total of 99 ares per farm, one of the lowest per household cultivated areas in Madagascar and soil degradation seem to be the most important constraints to higher agricultural production and, related, poverty alleviation. The average cultivated area per household masks very high variability between the Lac Alaotra zone and the Center and South Highland lower farm size of 70 ares.

The Mid West

This is the least densely populated agro-ecologial region, ranging from the Fivondronana of Ankazobe in the North to the Fivondronana of Ambalavao in the South. The farming systems are based on upland crops with rice cultivation being done in the narrow depression between the tanety. There is a diminishing gradient of rainfall from the North to the South, with a variation from 1,400 mm to 800 mm per year. Altitudes vary from 700 to 1,000 meters, gradually decreasing to the Western part. Soils are generally poor and acid, and prone to erosion. The Mid West depends on road transport, often in bad shape, for intra and inter-regional exchange. Rainfed rice dominates rice production. 42% of total cultivated area is occupied by rice. Upland crops, i.e. mainly cassava and maize, are the other main food crops.

This region is the most insecure region of Madagascar. 63% of the households declare to be living in a village where theft occurred in the last year. This might be linked to the remoteness of certain zones such as Maintirano, Kandreho, etc. One consequence of the remoteness and insecurity seems to be the low access to agricultural extension services: only 18% of the households report having access to it, half the national average. On the other hand, farmers have a relatively high level of education, with 69% household heads declaring to have finished primary school. Land rental markets are thin: only 6% of land is cultivated under a rental or sharecropping arrangement.

Determinants of Agricultural Production

Method and Description of Variables

In this section, the model and the variables used in empirical estimation are discussed. A more complete description of model and methods is given in Annex 2. First, the dependent variable is discussed and, consequently, an overview of the variables and their anticipated effect on agricultural production is given. These variables include land, labor use, capital, institutions, and natural shocks shows the descriptive statistics of the variables that are used for regression analysis.

To assess the determinants of farm production, we opt for the use of a primal production function approach. Due to data restrictions, particularly the non-existence of sub-regional data on input prices and agricultural wages in the EPM of 1993, the primal approach of production function analysis seems more appropriate than a profit function analysis. The use of Total Value Product as a proxy for production is imposed by data availability. The use of TVP, instead of physical quantities, as dependent variable in a production function analysis might create extra problems. First, as the TVP includes unit prices and quantities, interpretation becomes more difficult. Second, the dependent variable is sensitive to errors in more variables prices and quantities, so the error term might be larger. Third, aggregation problems as well are likely to affect the estimation, again tending to enlarge the error term. Different tests were done to select the appropriate model. The results favor the square root generalized Leontief function. The advantage of this specification is that the elasticity is flexible across the sample. In order to get at possible different returns of determinants, an evaluation of this elasticity and the marginal product is done at the means of every expenditure quintile.

Land

Four variables related to land and land quality are used in the regression analysis: cultivated land by the household, the percentage of titled land, the area under rental agreement, and irrigated area. It is expected that the more land and the higher the quality of the land, the higher agricultural production. The literature is ambiguous on the effect of land tenure on productivity. The common wisdom is that temporary or insecure property rights might have a negative effect on agricultural productivity, as there is less incentive for investments by the temporary or insecure owner. However, empirical results on land titling are mixed. Studies for different countries in Central Africa show neutral effects of legal land rights on agricultural productivity. Other studies have documented that insecure rights lead to less input use, investments, and, therefore, lower productivity. We expect the argument for secure property rights to lead to a positive coefficient for titling and a negative effect for the rental land variable. On the other hand, there is a potential second, positive, effect of rental markets. Credit constrained owners might rent out to tenants who have better access to liquidity, labor and inputs.

Conclusions

This report analyzes the link between agricultural production and poverty. To this effect, we rely on primary data analysis of the agricultural data of the national household survey. The major implications of this quantitative analysis for the agricultural sector, given the technologies that are currently available to the Malagasy farmers, are summarized below.

Education is an important determinant of agricultural productivity. The results of the regression analysis indicate that households where the household head finished at least primary education show 8 % higher agricultural production compared to households where the head did not finish primary school. Additional secondary education does not show a significant effect on agricultural production. Moreover, results indicate that the poorest quintile benefits disproportional compared to the richest quintile of primary education: 11% compared to 6% respectively.

Diversification in high value crops contributes to increased welfare. Farmers who grow export crops, vegetables or fruits are represented more in the richest quintile than in the poorest quintile. Poorer households depend relatively more on upland crops for their agricultural income than the richer households do. One "less valuable" upland crop that should be mentioned is cassava. It has been shown in other studies that cassava has the characteristics of an inferior crop, mainly grown and eaten by the poor. Hence, any success in improved productivity and consequent lower prices for this crop would be of direct benefit to the poorer part of the population.

The presence of legal agricultural land titles is shown to have small benefits for productivity at the national level. While an improvement of secure property rights that can be established in a cheap way might benefit agricultural productivity and efficiency in some regional settings, increasing attention should be paid to include the poor in this titling process. It is shown that the rich hold relatively more secure titles to land than do the poor and some studies have shown that the rich might even get titles to the detriment of the poor.

Rental agreements are shown to have a positive effect on efficiency and on equity. These results suggest that more secure rental contracts might be beneficial for poverty alleviation and for the agricultural sector. Two types of incentives seem to drive rental or sharecropping contracts. On the one hand, rich households rent out to smaller farms as they can not efficiently cultivate the extra land they possess. On the other hand, poorer households that do not have the means to cultivate the land might prefer renting out land to richer households. It seems that in both cases rental agreements are used to the benefit of the poorer households. Therefore, the Malagasy law that still prohibits sharecropping arrangements seems obsolete. Given that land sales markets are thin and will be very difficult to activate in the Malagasy context in the short run, a better pro-poor policy would be to ensure a more secure contractual environment for rental agreements.

On average, poorer households show low labor productivity while returns to a unit of extra land for them are large. Hence, it would make economic sense to assure better access to land for the poor. This can be achieved through more active land markets as well as increased mobility within the country as there are still fertile and productive areas in Madagascar that are little used due to insecurity or lack of infrastructure. Labor markets seem to function imperfectly in Madagascar as shown by the large impact of the family composition on agricultural productivity, especially for the poorer households. Given that results seem to indicate that the poor depend heavily on agricultural wage labor, they would probably be the biggest beneficiaries to increased mobility. This mobility could potentially be increased through a reduction in rural insecurity and through the development of better rural infrastructure. While we could not test for the effect of rural road infrastructure directly, it is widely assumed that it is a major constraint for improved agricultural productivity. Labor intensive construction of rural infrastructure might benefit the poorer part of the population through increased employment opportunities with the additional benefit of creating better functioning input and output markets, leading to higher agricultural productivity.

Modern input use shows to have high pay-offs, especially for the poor. This suggests that an improvement in the functioning of credit market might be an important way to solve the problem of insufficiency of cash at the beginning of the agricultural season. However, it seems that targeting of credit to the poor is difficult due to lack of collateral. Hence, they rely disproportional on informal credit systems that charge significantly higher interest rates. A second policy option is to encourage non-farm or off-season activity for improved agricultural production. A third way to improve input use is through stimulation and better enforceability of vertical contract agreements. This might shift the price and production risk towards firms, which have a higher risk bearing capacity. Such success stories in Madagascar in the agricultural sector exist for barley and tobacco.

Some important variables for agricultural production could not be handled in this analysis. First, agricultural extension services might contribute to higher agricultural productivity as it might improve the information on as well as access to new technologies. However, evaluating the effect of agricultural extension with the data at hand is difficult because of the heterogeneity of extension services, lack of information on agricultural extension in the dataset, and the problem of endogeneity. Second, the effect of road infrastructure on agricultural production was also not considered. It is evident that better transport infrastructure leads to higher output and lower input prices and creates therefore incentives for higher agricultural production. However, evaluating the effect of access to roads is difficult as roads are clearly endogenous variables and as one would need to consider the whole road network, not only the presence of a road in the village. It seems that this effect can only measured through panel data. Lastly, health services might affect agricultural production. However, its effect is difficult to assess with our data due to endogeneity issues as poor health results in low production but causality runs also the other way as low production, low consumption and low income affects the health situation of the household.

Finally, data limitations were the driving force for the choice of the approach used in this analysis. The major limitations of this approach can be summarized as follows. First, as production factors were not separated by crop or plot, we were obliged to aggregate over all crops. Separate analysis of lowland and upland or by crop would give better estimation on the benefits of, particularly, irrigation, agricultural expenditures, and extension services. Second, as these are cross-sectional data, endogeneity problems seem to exist. Panel data or good instruments would be needed to achieve better estimates of the different determinants. Third, having data on input and output prices and wage rates would allow to establish the link agricultural supply and market prices, and would allow for profit analysis in which case there might be less endogeneity problems.

Given that a new household survey, with an extensive agricultural section, is to be implemented by the end of this year, crop specific analysis might be possible, which would allow for better priority setting in the agricultural sector.

Chapter 5

The Structure of Public Expenditures, Agricultural Income and Rural Poverty

The provision of public goods and the related issue of ameliorating the impact of market failures are the classical justifications for government intervention in the economy. In reality governments do not do a good job in fulfilling these roles. Governments intervene in markets that cannot be even remotely considered affected by failure or imperfections. A large share of the financial, human and institutional public resources is spent in private goods, not in public goods. Misguided public policies that interfere with essentially efficient markets have been for a long time a central concern to economists. The literature documenting and measuring the efficiency costs of market interferences for many commodities and many countries around the world is indeed massive. On the basis of these studies international organizations as well as several other economic institutions have forcefully argued for the removal of government interference with markets operation for several decades.

The high degree of attention that the economics literature has given to issue contrasts with the relatively little concern that issue has received from both academic and policy economists. This despite the potentially large efficiency and equity losses arising from misguided allocations of public expenditures. The fact that raising government revenues is costly and that the availability of government financial, human and institutional resources is usually tight, implies that deviating part of these resources to private goods crowds out the supply of public goods. Spending large volumes of public resources in private goods means that governments are over taxing the private sector, or under supplying public goods or both. Whatever the means used by the public sector to finance spending in private goods, it implies efficiency losses associated with Education and health care though not pure public goods have the inevitable distortions caused by raising more taxes than needed or by under providing public goods.

The under provision of public goods may have serious consequences for the productivity of private investments as public goods are important complementary assets with private capital. A low supply of public goods means scarcity of human capital, under investment in research and development less infrastructure, insufficient environmental protection, etc. All these are important assets that contribute to increase the productivity of private investments. Hence their scarcity is likely to constitute an obstacle to economic expansion.

The biased structure of public expenditure allocation too much private goods and too little public goods may have important equity implications as well. For reasons to be discussed below, government expenditures in non-social subsidies and other private goods tend to be directed to the wealthier segments of society to the detriment of the poor. Over taxation that spending in private goods may cause, also has negative equity consequences. It is well known that in most developing countries taxes are highly regressive especially because of its great reliance on indirect taxation. So the additional taxes needed to finance subsidies to the wealthy are in a significant part paid by the poor.

In addition, the under supply of public goods affect the welfare of the poor much more than the welfare of the wealthy. The only significant assets of the poor are their labor force and, for the rural poor, their natural resources. The income of the poor is thus extremely dependent on their ability to enhance their human capital and to protect the natural assets, both important public or semipublic goods. There have been many empirical studies that have measured the impact of specific public goods such as R&D, rural roads and other infrastructure on agriculture as well as on other sectors. What seems to be missing, however, is a broader approach that may permit evaluating the impact of public goods vis-à-vis other types of government expenditures on efficiency and social equity. Without such a comparison the conceptual And at the same time applauding or guarding accomplice silence a and policy implications of findings pointing to the high pay-offs of certain specific public expenditures is in part obscured.

At the policy level, there seems to be a degree of consensus regarding the importance of an adequate provision of public goods for economic expansion and social equity. However, the same lack of a clear conceptual framework and the general weakness of empirical studies due in part to such a conceptual deficit also mean a certain weakness in the policy recommendations. It is clearly a rather empty policy statement to advice governments to increase expenditures in education, health, roads etc. without pinpointing the source of the low expenditures in such goods in the first place and without indicating what expenditures to cut or what taxes to raise in the second place. In any case, the concerns shown by the profession and international institutions about issue appears to be a pittance compared to the overwhelming focus in reducing government market interventions. Below we provide evidence suggesting that such lack of balance is not justified.

In most countries the performance of agriculture and the evolution of rural poverty are likely to be heavily dependent on economic interdependencies with the rest of the economy and with the rest of the world. Much of the direct interactions occur through mechanisms associated with government market interventions and with the level and composition of public expenditures in the rural sector discussed earlier. In addition, both agricultural productivity and rural poverty may be affected by the performance of the non-rural economy as recently hypothesized by Gardner. A dynamic non-rural economy may have positive spillover effects on the rural sector and on agriculture: Higher commodity prices, greater availability of credit, more new technologies, and so on, are all factors that promote a faster expansion of agriculture. In addition, better employment opportunities and higher economy-wide wages are likely to also have positive influences on rural poverty through both migration and remittances from the urban sectors.

The objective of this paper is to measure the quantitative importance of the above factors on the performance of agriculture and the evolution of rural poverty for a subset of countries in Latin America. In particular, we are interested in shedding light on the following questions: What would be the impact of changing the structure of rural public expenditures e.g., reducing expenditures in private goods and concomitantly increasing public good expenditures while keeping total expenditures constant on agricultural income and poverty? How does the impact of such policies compare with the effect of traditional policy prescriptions associated with reducing certain market interventions? Is increasing public expenditure in the rural sector the main mechanism by which the performance of the non-rural sector affects agriculture?

A brief conceptual framework and hypotheses are first presented. Next we review existing empirical evidence suggesting that the issues raised are likely to be of great policy relevance. Then we present a quantitative description of the evolution of nonagriculture factors likely to affect the performance of agriculture over the period 1985- 2000 for ten Latin American countries. The countries considered have consistent rural public expenditure series with a degree of detail sufficient to allow us to aggregate them into expenditures in public goods and in private goods. We show that for the countries considered a large share of total public expenditures in the rural sector is constituted by private goods. Next we discuss the econometric model, and in the following section the econometric results are presented. We conclude with an evaluation of the empirical findings as a contribution to the development of a policy agenda that is broader than the conventional one.

Conceptual Issues

The central focus of this paper is to study how factors external to agriculture impinge upon its performance and how such factors have affected rural poverty. We emphasize one mechanism that having potentially a large policy connotation has, in our view, been insufficiently emphasized and analyzed by the literature: The way in which the public sector spends tax revenues in the rural sector. We do not intend to evaluate whether or not governments spend enough in agriculture. This is an old debate. We instead are interested on the *structure* or composition of government expenditures in rural areas including of course in agriculture. But we are not interested either on analyzing the impact of particular government expenditure items. The literature is full of studies that do this: Measuring the impact of road expenditures, R&D investments, credit subsidies, farm extension, rural education, expenditures in rural health, are some of the favorites in the literature. This literature makes a contribution by showing that certain specific items may have high pay-offs while others not so high. These findings may have relevance by emphasizing the need to improve the cost-benefit evaluation of particular public expenditures and investments. But it is difficult to obtain more general policy implications from them.

We instead use a broader taxonomy of public expenditures that is conceptually and policy grounded. The classification of public expenditures into *public* goods and *private* goods arises from the conceptual view that governments' prime responsibility is to use tax-payers money in supplying public goods, including goods and services that are under supplied as a consequence of missing markets or of market imperfections. To be sure there are many grey areas in this respect. Sometimes expenditures in private goods are difficult to distinguish from expenditures in public goods. But by and large these are the exception: in practice more often than not the separation is quite sharp.

Political motivations and socio-economic consequences of public expenditures.

The source and political motivation of the government's decision to invest in public or private goods is fundamentally different and such different motivations cause important socio-economic effects of the public private good choice. By definition the benefits of public or semi public goods tend to disperse much more among the population than those of private goods. This implies that the political lobby targets much more private goods that wholly benefit particular lobby groups than public goods of which the lobby group shares only a fraction of the benefits. The struggle by lobby groups to elicit subsidies from the government is often their prime motivation. It is well known that most successful lobbies in developing countries are typically small groups with a clear common interest and objectives. In addition, and most importantly, successful lobby groups almost always have enough financial means to bribe and to pay political campaign contributions to politicians, or have enough social status to be able to affect public opinion through a variety of means.

Subsidies may, however, be directed in some cases to mitigate mar 5 ket imperfections and externalities that otherwise may prevent even wealthy entrepreneurs from optimally invest. The evidence available, however, seems to indicate that subsidies that can be justified in this way are rarely implemented. The above requirements for constituting a successful lobby group are often satisfied by economic elites and rarely by the poor. The implication is by now a stylized fact: Government expenditures in subsidies and other private goods go almost always to the wealthy, only very rarely to the poor. This is in sharp contrast with expenditures in public goods which generally produce benefits that tend to disperse across all groups in society, poor and non-poor. The social equity effects of the composition of public expenditures are obvious: The greater the share of government expenditures in private goods the worse, ceteris paribus, is income distribution and poverty. Because the structure of public expenditures tends to be more biased towards private goods when the economic power of the elites is in the first place already greater, such biased composition of public expenditures is both cause and effect of poverty and social inequity.

The structure of public expenditures and economic efficiency. The public private good dichotomy is grounded on another factor which has to do more directly with economic efficiency than with equity. Investment in public goods provides factors of production that are almost by definition rarely supplied by the private sector. For this reason one can reasonably expect that such investment will fill a vacuum that is not fulfilled by the private sector. As a consequence, a complementary relationship between private investments and public goods is likely to arise. Government expenditures in private goods, by contrast, finance assets that can equally be financed directly by private entrepreneurs. Even more importantly, as argued above these government subsidies go to the wealthy, those that face generally no constraints to invest at optimal levels that are likely to face no liquidity constraints or other types of market imperfections. As shown below, the effect of these subsidies is almost always detrimental for economic efficiency and, far from contributing to promote private investment, they tend to crowd it out and to increase the consumption of the wealthy instead.

Thus we have another important contrast that justifies the public private goods dichotomy: While increasing public goods is likely to promote economic growth directly as factors of production and indirectly through its positive effect on private To be sure, there exists the option of increasing taxes so 6 that at least the financial resources for public good provision at adequate levels may still be available. But raising taxes is costly in terms of economic efficiency and also entails political costs to the government. Moreover, more tax revenues may allow having the needed human and institutional capital required to administer the provision of public goods only in the long run.

The crowding out of public good expenditures within the government's resource budget can have large negative effects on social welfare. Governments forego investments in public goods that according to many studies have extremely large rates of return. The survey by Alston et.al. reviews hundreds of studies around the world estimating rates of return to agriculture R&D revealing extremely high and non-declining rates. Psacharopolous's survey does the same for studies measuring returns to education over many countries, also showing very high rates. Similarly several studies at IFPRI by Fan and his associates have found unusually high returns to a variety of public infrastructure in agriculture in India and China. The World Bank has gathered evidence pointing to also very high rates of return to investments in certain investment in protecting the environment. Thus, on the basis of this rather massive evidence one may presume that governments under invest in public goods the persistence of enormous rates of return can only be explained by insufficient investment and that such under investment is likely to have large negative effects on welfare. This conclusion is reinforced by further empirical evidence showing that subsidies have low Suppose the subsidy is 50% of the value of the investment cost and the rate of return per annum of the investment is quite high, 20%. Assume further that the producer estimates that the probability of getting the subsidy next year is 0.5. If the producer is risk neutral he she will decide to delay the investment for one year. or even negative economic rates of return and that they crowd out private investment. We turn to these issues next.

Subsidies may *directly* crowd out private investments. Consider what is often regarded as a "desirable" subsidy; the government offers to pay a portion the costs of a particular investment. Assume further the best possible circumstance in terms of the allocation of the subsidy: The subsidy is of course rationed as the funds are obviously less than the demand, but their allocation among producers is transparent, not subject to corruption. Consider an investor that is able to extract a profitable return out of an investment even in the absence of the subsidy that potentially may qualify for the subsidy. Suppose that in that year there was a large demand for the subsidy and that the investor was not lucky enough tc get the subsidy. The producer may go ahead with the investment and never get the subsidy anyways since it is a profitable one. Alternatively, she he may opt to postpone the investment and try again next year in the hope of then getting the subsidy. If the expected value of the subsidy is sufficiently large to compensate the foregone profits in one year the producer may decide to delay the investment. Thus, investments that are privately profitable may be postponed as a consequence of the existence of the subsidy.

Among the investors that actually got the subsidy there are two types: Those that would have implemented the investment anyways and those that would have not because they would not be able to get high enough returns but they in fact invest because they got the subsidy. For the former the subsidy was ineffective, the subsidy is likely to promote more consumption by them rather than more investment as intended. For the latter the subsidy was effective in causing them to invest but at a low social return. Thus, the subsidy scheme does two things, it increases consumption of some producers and it causes a reallocation of investment from producers that are able to obtain high private and social rates of return to the investment to producers that obtain a low social return.

The above example is not just a curiosity. It illustrates a phenomenon that has received important empirical support in recent years. That subsidies, at least in the form in which they are usually allocated, do not generally promote investment or more R&D has been shown by several studies in various countries. Empirical studies using detailed firm level data by Bregman et.al. for Israel, Fakin for Poland, Lee for Korea, Bergstrom for Sweden, Estache and Gasper for Brazil, Harris for Ireland and several others have shown that subsidies and corporate tax concessions are at best ineffective to promote investment and technological adoption and, in some instances, even counterproductive. Crowding out of private investment as a consequence of the subsidies occurs.

Subsidies may *indirectly* crowd out private investment in the intermediate run. The third type of crowding-out of private investment due to public expenditure biases toward subsidies is long- run in nature. The low stocks of public goods caused by continuous under investment in public goods cause low productivity of private investments over the long-run. This, in turn, is translated into slower investment and slower productivity growth.

The structure of Public expenditure and rural poverty. Public goods affect rural poverty through a *direct* effect and through an *indirect* effect:

(i) Direct effects. Important components of rural public goods are human capital and investment in the protection of natural resources and the environment. Human capital is the main productive asset for most poor. However, as a consequence of credit and other market failures they are not in general able to fully finance investments in human capital, regardless of how high the rate of return to these investments might be. They are, therefore, largely dependent on the public sector as a source of financing for these investments. On the other hand the rural poor are also highly dependent on natural resources as a source of subsistence. Resource degradation is often disproportionably paid by the rural poor. Increasing investment to The assumption of constant returns to scale in agriculture has 11 recently been often not rejected by empirical studies using farm household data. See, for example, López and Valdés who provide estimates for five countries in Latin America. protect natural resources and in mitigating environmental externalities is, therefore, vital to reduce rural poverty.

(ii) Indirect effects. As discussed earlier an adequate stock of rural public goods contributes to sustain growth for agriculture and related rural industries. In general much of the rural industry linked to agriculture is highly intensive in unskilled labor. Moreover, it appears that the rural sector in many developing countries is important enough within the unskilled labor market to significantly affect the real wages for unskilled workers even at a national level. Thus a dynamic rural productive sector is a vital source of employment for unskilled workers as may play a role in supporting higher real wages. To the extent that most poor are unskilled workers, these effects may be important in reducing poverty.

The Data

In this section we examine a new data set for rural public expenditures for ten countries in Latin America recently assembled by FAO Regional Office with the support of the World Bank and others. The annual data for the ten countries covers the period 1985-200012. An advantage of this data is that is quite disaggregated while at the same time it has been measured using a reasonably consistent methodology across countries and over time. In addition the public expenditure data as provided is quite comprehensive, covering the vast majority of the variety of items in which the public sector spends money in the rural areas. The level of detail of the data allows us to obtain a fairly accurate measure of the evolution of total government expenditures in rural areas over the period.

The data for public expenditures in the rural sector covers all rural expenditures, including expenditures for agriculture as well as rural infrastructure, animal and plant sanitary protection, environmental expenditures and social services to rural areas. The latter group covers expenditures in rural education, health and other social sectors. All in all rural public expenditures are disaggregated into 12 categories. The expenditure groups including in brackets their respective average country share in total public expenditures for the year 1995 are the following: marketing assistance irrigation technological generation and transfers, soil conservation, forest promotion, plant and animal sanitation, communications and information "focalized" expenditures directed to particular commodities integral rural development rural public infrastructure rural social services and others.

To allocate the above expenditures into private versus public goods we first account for items that clearly fall into one or the other category. Next we deal with those categories that fall into a grey area where it is difficult to determine a priori whether they correspond to public or private goods. In these cases we have looked into more detailed components falling into these categories in the various country statistics available. This allowed us to obtain an educated guess as of what proportion of these expenditure items should be approximately allocated to public or private goods. Finally there are some minor items for which we were not even able to have a reasonable guess. In the latter cases we simply use a 50-50 rule.

Among the items that clearly qualified as public or semi public goods were: Technology generation and transfers, soil conservation, plant and animal sanitary protection, communications and information services, rural infrastructure, and social services (education, health and anti-poverty programs). In classifying the above expenditures as public goods, we considered the fact that they include areas where there are important positive externalities associated with such services technology generation and diffusion, soil conservation, plant and animal sanitation that were expenditures directed to at least in part palliate the impact of market failures social services and expenditures in expanding assets that directly important features of general public goods communications and public infrastructure.

One may argue that classifying irrigation expenditures as private go 13 ods while other public infrastructure such as roads, bridges, etc as public good may appear as arbitrary. However, unlike other forms of public infrastructure irrigation projects are targeted to a limited number of farmers who usually pay little if at all for the water service, *excluding* everyone else from directly benefiting out of the irrigation investment.

Items that were considered private goods are: Commodity-specific or focalized expenditures, marketing promotion (excluding expenditures in communications and market information which are among the public goods items) and irrigation expenditures. An important portion of "market promotion" includes direct subsidies to producers as well as subsidized credit. Similarly, most of the irrigation services are completely or, almost completely subsidized. The third group, included the categories forest promotion, IRD programs and "others". Among the latter three items, a high proportion of the forest promotion expenditures were in the end allocated to public goods while a smaller fraction of IRD were considered to qualify as public goods. An important portion of tree planting expenditures is allocated to eroded areas, thus inducing positive externalities by protecting soils, contributing to regulate water regimes, etc. IRD expenditures are mostly subsidies to promote particular patterns of agricultural expansion that the planners desire. As such a high portion of them does not qualify as public goods.

The Results

We estimate the reduced-form for per capita agricultural GDP using various methods of estimations and aggregating the data in several alternative ways. Our interest is primarily probing the robustness of the estimated coefficients. Here we present two types of data aggregation. (1) Regressions using country data aggregated for relatively long periods of time, in the case presented in the ensuing tables five year averages; (2) regressions using annual data with various specifications for the lag effects. *"Long-run" effects.* An attractive feature of (1) is that is consistent with the fact that much of the effects of the independent variables upon per capita agriculture GDP are likely to be embodied through processes that may take considerably more than one year to have an effect. One can of course specify dynamic processes and then design a lag structure accordingly. This is obviously difficult to do. So working with several year averages instead of annual data might help in elucidating what can be considered longrun effects of the explanatory variables. A drawback of the approach is that one loses degrees of freedom by reducing the number of observations. We consider several aggregation options using simple averages. In general as we increase the length of the observation period the estimates show a fairly consistent pattern: their absolute values tend to increase while their degree of significance changes only moderately as we increase the length of the observation periods.

And this is obtained *without* increasing the total government expenditures. By contrast, increasing government expenditures without altering its composition is much less effective in raising per capita agriculture incomes: A ten per cent expansion of government outlays causes on average only a 0.6% increase of agriculture income. In 1996-2000 the 50% of the countries that spent the least in rural areas spent about $35 per capita while the top 50% of the countries spent about $74 per capita. Thus if the average bottom spender increased their per capita outlays to levels comparable to the average top, per capita agricultural GDP in that country could increase by more than 3%. With respect to the share of subsidies, in 1996-2000 it ranged from approximately 30% average for the bottom half to 65% for the top half. This means that if an average country in the top half could readjust its public expenditure share to the level of the average country in the bottom half its agricultural GDP may increase, ceteris paribus, by a whopping 12.5%.

The results confirm the positive role of trade liberalization on agriculture income. This reflects the fact that agriculture is still discriminated against through protective trade policies. An opening up of trade policies consequently is not only good for overall economic efficiency but also for agriculture income. The marginal effect of trade openness index on agricultural per capita income is about 0.002 and statistically significant. Moreover, this estimate is remarkably stable and robust to changes of specification. To get an idea of the meaning of this number it is useful to indicate that the trade openness index for the period of the most open country within the sample is about 60 points higher than that of the country that is least open. This means that if Uruguay had been as open as Costa Rica over the period its per capita agricultural income would have been about 0.12 % higher. Thus, though the effect of trade openness is highly significant, its quantitative value is quite small, a result that is consistent with other studies showing that the static gains of trade liberalization may not be too great.

The estimates are consistent with an important positive effect of the rest of the economy upon agriculture. This large effect may, however, be affected by biases due to omitted variables that could exert similar effects upon agricultural and non-agricultural GDP. The FE estimators do control for the impact of fixed omitted variables that are country specific. But not for shocks that happen over time and that could influence the dependent variable and be correlated with the explanatory variable. We have also used instruments which could at least deal with reverse causality biases, but certainly the issue of time varying omitted variables remains. In any case, it appears that the performance of the non-agricultural sector positively affect agriculture. This could be due to a Both the FE and RE estimates passed the White's test of heteroscedasticity neighborhood effect; when the overall economy flourishes, investors would be more willing to invest in all sectors.

The fact that now we have much more observations allow us to use annual time dummy variables to control for unobserved shocks common to the countries that vary over time in addition to the random or fixed country effects. The pattern of significance of the estimated coefficients and their sign are highly consistent with the "long-run" estimates reported earlier. Consistent with the short-run long-run interpretation, in general the elasticity estimates using annual data have lower absolute values than the coefficients estimated using the five-year averages. In particular, the effects of total government expenditures in rural areas are still positive and statistically significant, but their values are about 50% lower than the values estimated using five-year averages. Similarly, the share of government subsidies in total expenditures is now negative and significant as before but the estimates are smaller in absolute value. Still, the message is the same: total government expenditures are important in promoting agricultural income, but the structure of public expenditures is even more important. Switching resources from subsidies to public goods can be an extraordinarily effective instrument to achieve faster agricultural growth.

The reduction of the size of the effect of the non-agriculture income is even more pronounced than that of the public expenditures variables. In fact, the effect of nonagriculture GDP falls by about 66% when annual data is used compared to five-year averages, though the coefficient is still significant. This drastic reduction of the size of the coefficient may reflect not only the short run long run dichotomy, but also a possible reduction of the size of the omitted variable biases caused by the fact that we now control for the effects of omitted variables that do change over time. In any case, an elasticity of about 0.13 still reflects a sizable positive impact of the rest of the economy upon agriculture.

Rural poverty. The quantitative analysis of the impact of the level and structure of public expenditures on rural poverty is hampered by insufficient availability of rural poverty estimates for the countries considered and especially by issues of data quality and comparability discussed earlier in the paper. These data problems are reflected in the econometric estimates which are not nearly as robust and clear cut as the estimates of agriculture GDP reported earlier reports estimates of a rural poverty equation using data from the 1980s as well as 1990s. Because the data for the earlier years seems to have more problems than that in the later we also provide estimates using only the data for the period 1992-2000. The fact that we have a rather shallow panel, with most countries having just two observations throughout the period, implies that the FE estimates are highly unstable and tend to yield implausible estimates. For this reason we present here only the RE estimates which tend to be more robust and stable. In general we are not able to detect any direct effect of the structure of public expenditures on rural poverty. The estimates of the effect of the share of subsidies are positive and significant in some selective runs but by and large are not statistically significant. The same is true for the effect of total public expenditures. It is interesting to say that when we do not control for the composition of public expenditures, the estimates for total expenditures tend to be positive. This might be consistent with the hypothesis that, regardless of the public private good composition, public expenditures tend to be socially regressive. That is, even the public goods provided may not be primarily those that benefit the rural poor.

The most robust finding is the negative effect of agricultural GDP on rural poverty. Non-Agriculture GDP also has a negative effect on rural poverty, but the size of its effect is more modest. This is of course not surprising as the effect of the rest of the economy on rural poverty is more indirect than the impact of agriculture. In any case, the elasticities of agriculture and non-agriculture are both statistically less than one. By contrast, studies relating poverty to overall GDP per capita often yield elastic effects. Though we are not able to capture any direct effect of public expenditures on rural poverty, the fact that the level and structure of public expenditures have important effects on agriculture per capita income suggests that the indirect effects are significant. In particular, a 10% reduction of the share of subsidies in total expenditures may cause an increase in agriculture per capita GDP of about 2.3% which, in turn, may induce a 1.2% reduction in rural poverty.

Conclusions

This paper has provided evidence suggesting that the structure of public expenditures is an important factor of economic development. The quantitative importance of this factor appears to be greater than the traditional factors on which the development literature focuses.

The econometric results provide estimates that suggest that governments over spend in mostly politically-motivated private goods and subsidies and under spend in public goods and anti-poverty programs. Though the very large estimated effects of expenditure composition may be surprising to some, one should note that these estimates are highly consistent with two previous pieces of evidence available in the literature. First, the literature reports extremely large rates of return to many forms of public goods often 25% to 60% per annum and above including R&D, education, health care, some infrastructure, certain environmental protection investments, and others. Also, those studies that have performed analyses over time generally find that the rates of return show no tendency to decline over time. Recent surveys of this literature, cited earlier in the text, find an amazing degree of consensus on the estimated rates of return obtained in many studies that evaluate rates of return to public goods around the world. The fact that the studies reviewed have used different approaches, diverse data bases and have been applied to a large number of countries around the world, means that this consensus should be taken seriously. These extremely high and non-declining rates of return suggest that there are important investment opportunities in public goods that are not being fully exploited. That is, it suggests under spending in public goods.

Second, many firm level studies around the world some cited earlier in the text have found that subsidies aimed at promoting investment, employment, and growth do not achieve these objectives. In some cases they are counterproductive to these goals. Thus, while the rates of return to public goods are very high, the rates of return to private goods or subsidies are very low or even negative. Yet, governments tend to spend a large share of their resources in the goods that have low returns while at the same time forego investments with very large rates of return. Switch spending priorities and you can obtain massive dividends. This is what the econometric results are saying.

As discussed earlier in the paper, governments make such an aberrant choice mostly because they respond to a variety of politico-economic forces. These forces at play are not only associated with open government corruption practices, but they may also reflect a highly unbalanced configuration of political forces. The elites, because of their wealth and "contacts", are able to influence even well-meaning governments to their own advantage even if the rest of society suffer loses. Apart from bribes and political campaign contributions, the elites are able to create an ideological climate, mainly through the media and intellectual circles that they control, that favors or accept such flagrant misallocation of public resources. International institutions can, if they want it, be important players in creating domestic conditions favoring correcting such misallocation. They can also use their influences without and within the countries to reduce rather than deepen the massive imbalances in lobbying abilities between the elites and the rest of the civil society.

Chapter 6

Rural Poverty: Old Challenges in New Contexts

Poverty is still a predominantly rural phenomenon. However, the context of rural poverty has been changing across the world, with high growth in some economies and stagnation in others. Furthermore, increased openness in many economies has affected the specific role of agricultural growth for rural poverty reduction. This paper revisits an 'old' questionhow does growth and poverty reduction come about if most of the poor live in rural areas, dependent on agriculture? What is the role of agricultural and rural development in this respect? Focusing on sub-Saharan Africa, we use theory and the available evidence to conclude that changing contexts has meant that agricultural growth is only in particular settings crucial as an engine for growth, more specifically in landlocked, resource poor countries, which are often also characterised by relatively low potential for agriculture. However, extensive market failures in key factor markets and likely spatial effects give a remaining crucial role for rural development policies, including focusing on agriculture, to assist the inclusion of the rural poor in growth and development. How to overcome these market failures remains nevertheless a key issue for further research and evaluation.

Introduction

Even though poverty has been coming down in some parts of the world in recent years, not least in East Asia and China, and more recently in South Asia, its persistence in large parts of Africa and elsewhere keeps it high on the agenda. Poverty persistence is to a large extent related to a poor growth performance in national economies. Furthermore, in large parts of the world, poverty also remains in sheer numbers still mainly a rural and agricultural phenomenon, in that most of the poor depend on the rural sector for its livelihood.

This paper will revisit some well-rehearsed but nevertheless relevant questions. What is the place of rural development and agricultural growth in growth and poverty reduction? What are the main rural constraints on growth and poverty? Has more recent economic theory and recent empirical research given more guidance about what can be done? These issues will be discussed keeping in mind some of the poorest parts of the world, most notably sub-Saharan Africa. We will use theory as our guide to structure our discussion, and discuss empirical evidence, even if incomplete, to back up the conclusions. These questions are well-rehearsed and feature prominently in many general discussions on development. Textbooks typically contain discussions at various level, for example, Ray or Bardhan and Udry. Longer treatments of these themes can be found in Timmer or De Janvry et al. What is different in this paper is that we want to revisit some of the theory and evidence specifically with sub-Saharan Africa in mind; as a result it is highly selective. Overall, the experience with agricultural growth, growth in general, and poverty reduction over the last decades has been disheartening in this region. It has provided the context for renewed calls for a strong focus on agriculture in Africa as a necessary condition for growth and poverty reduction. For example, Sachs has been most vocal in calling for a 'green revolution' in Africa as an essential part of its development strategy. More nuanced analyses, such as the recent World Development Report 2008 on Agriculture for Development, emphasize the crucial role of agriculture as being "vital for stimulating growth in other parts of the economy" in sub-Saharan Africa, requiring strong productivity increases in smallholder agriculture. This paper will revisit these arguments in the context of the substantial heterogeneity in circumstances and opportunities across sub-Saharan Africa. In the next section, we first briefly summarize the current evidence on the evolution of rural poverty, contrasting the African experience with other parts of the world. The evidence is consistent with a classic view that a move out of agriculture is correlated with overall poverty reduction. In the African context, neither a large scale poverty reduction nor a move out of the rural sector is occurring. Of course, this does not prove a causal link. Nor does it prove that a focus on agricultural growth is not warranted.

In the subsequent sections, we ask how theory as well as the available evidence can inform us on what role agricultural and rural development has to play in both growth and poverty reduction in Africa. In section 3, we offer a discussion of the relevance of a 'macro' view on the role of agriculture in growth and poverty reduction. This will require us to revisit some of the 'old' and seemingly unfashionable questions related to sectoral and urban-rural linkages, to better understand the role of the relevance of agricultural and rural development in the Africa setting. One of the key problems in this review is the relatively sparse evidence, so we rely largely on a simple but suggestive model of ruralurban linkages to make sense of this evidence. Combining this with recent approaches to the question of the scope for growth in Africa, emphasizing its heterogenetiy in opportunities allows us to identify those specific settings in which agricultural development is likely to be essential in stimulating growth and poverty reduction, as well as the nature and role of rural development in other settings. In particular, it is argued that the role of agriculture is likely to be very different in different settings, depending on whether a country can take advantage of manufacturing opportunities, whether it is natural resource dependent, or whether it is land-locked with few natural resources. It is argued that especially in the latter, a focus on agricultural growth may have to be essential if difficult route out of poverty.

In section 4, the strong assumptions regarding the nature of markets implied in the macro-view are complemented by a micro-view, emphasizing different market failures, and the possibility of poverty traps. We will focus on three examples of serious market failures, related to credit, risk and spatial effects, reviewing their theoretical impacts and the available evidence. These market failures, especially in circumstances where they may lead to poverty traps, bring rural and agricultural policies back to the fore, leading to suggestive policy conclusions on the role of agriculture and rural development.

Rural Poverty Patterns

Poverty is still predominantly a rural phenomenon. Pick a random poor person in the world, and the odds are that this person will be living and working in the rural areas as a farmer or agricultural worker. Even though the data are not without problems, the most recent estimates suggest that about 76 percent of the poor in the world live in rural areas, well above the overall population share living in rural areas which is 58 percent. Sub-Saharan Africa is no exception: while it has the highest poverty rate overall, rural poverty is about a quarter higher than urban poverty, with 65 percent of the population living in rural areas, and 70 percent of the poor are in rural areas. At current patterns of growth, poverty reduction and population growth, poverty is likely to remain a predominantly rural phenomonen for the next few decades.

Is this changing. The figures in Ravallion et al. also offer a suggestive insight in the recent patterns of urbanization of poverty over the period 1993 to 2002. Even though the urban poverty rate has marginally decreased in the world, the urban share of poverty has been increasing from about 19 to 24 percent, as the urban population has grown faster than the rural population, largely due to in-migration.2 There is considerable variation in this pattern as well. In sub-Saharan

Africa, a very marginal decrease in rural poverty and stagnating urban poverty rates with a growing share of the urban population in the total population, is responsible for a more substantial urbanization of poverty and little change in total poverty. The global overall urbanization of poverty in the context of relatively substantial poverty reduction, and with the larger part of the 'stock' of the poor living in rural areas also implies that rural poverty has contributed most to overall declines in poverty: using a simple decomposition, Ravallion et al. calculate that about 80 percent of aggregate poverty declines stem from rural poverty declines. But this obviously does not prove any causality between, say, urbanization and poverty, or indeed that what happens in the rural or agricultural economy is in itself the cause of poverty declines.

In particular, these patterns have to be seen in the broader context of the economy. Growth in GDP per capita and poverty reduction are well known to be coinciding with a gradual decrease of the share of GDP from agriculture as well as the share of the population engaged in agriculture. For example, across all low-income countries, growth stood at about 5 percent per year and the share of agriculture has declined from 32 to 23 percent of GDP between 1990 and 2004.

Contrasting the experience in the 1990s of sub-Saharan Africa with the relatively fast growing regions in East Asia and Pacific, and South Asia, where substantial poverty reduction is taking place, is instructive. We observe growth of overall GDP far outpacing agricultural GDP growth in both East Asia and South Asia, but not so in sub-Saharan Africa, where overall growth and growth in agricultural GDP were more similar at 2.5 and 3.3 percent per year respectively.3 With population growth still at about 2.3 percent per year in sub-Saharan Africa, much higher than in these other regions, per capita growth has been minimal. In other words, there is little sign of a structural transformation in SSA despite signs of some urbanization of poverty, compared to the patterns observed in more successful regions.

For some, agricultural growth is seen as the key engine of growth. The higher recent growth rate in African agriculture is then a sign of hope, even if at current rates they can hardly be viewed as evidence of rapid agricultural transformation. In such a view, the key is to raise agricultural productivity to allow growth to take off. A standard conceptualisation of this process focuses on the existence of linkages in inputs and outputs between agriculture and other sectors. Showing sympathy with this view, in the World Development Report 2008, it is argued that in currently agriculture-based economies, high growth in agriculture may well the road to take-off. Even though the historical evidence on the role of agriculture in allowing the Industrial Revolution to start in Europe is suggestive, and agricultural growth was also significant element in the growth success of East Asia and China, it is much harder to argue that agricultural growth is *essential* to allow growth to take-off, and the evidence is at best only suggestive.

There is stronger evidence in favour of the view that agricultural growth helps growth to be more pro-poor, albeit possibly dependent on context. For example, in China, it is estimated that growth from agricultural growth contributed up to four times more to poverty reduction, than growth from industry or the service sector. However, the rather favourable land distribution may well have played a key role in this, a factor contributing to similar evidence in other East-Asian settings, such as Vietnam. Evidence from India suggests a more subtle message: the impact of the growth in agriculture is matched by growth in services in terms of their poverty-reducing effects, although these poverty-reducing effects from nonfarm growth are larger in states with higher initial farm productivity. Furthermore, evidence from Foster and Rosenzweig suggests that areas in India with the slowest *growth* in agricultural productivity had the largest growth in the rural nonfarm tradable sector.

This discussion of patterns and change in rural poverty helps to movitate our further analysis on Africa. It is clear that poverty is highest in rural areas, but is that enough to pay central attention to rural areas and to agriculture? Successful poverty reduction is not simply equated with relatively high growth in agriculture. At best, during periods of rapid poverty reduction correlated with growth, rural growth is likely to be important for poverty reduction, but successful growth is associated with growth in the non-agricultural sector fast outpacing agricultural growth.

In any case, it implies that understanding rural poverty changes cannot naïvely focus only on what happens in the rural sector, or in agriculture. It will have to be done in the context of overall growth and changes, taking into account rural-urban linkages. Arguably, the first systematic treatment of these issues are in the Lewis-model. This model is effectively part of a *theory of urban and rural interactions*, albeit with particular assumptions about the functioning of markets in urban settings, and especially, the nature of incentives and decision-making. Much work refined this type of analysis further, with a shared underlying research question: how does growth and poverty reduction come about if much activity and labour is initially in the agricultural or rural sector?5 These questions have somewhat been forgotten in much recent research. However, the context of the rural sector has also rather dramatically changed in the developing world, say compared to the 1970s and 1980s. In most of the developing world, not least in sub-Saharan Africa, market-based reforms "getting the prices right" has moved forward dramatically. In Africa, the context of farming and agricultural markets has substantially changed, and virtually everywhere important liberalization, mostly domestic but also increasingly of international trade, has taking place.

Globalization and increased openness, as well as investment in commercial agriculture and marketing, are changing the context slowly but in an irreversible way. This provides a different context to ask again. Can and should agriculture be the engine for growth. Can and should it be an engine for poverty reduction in the context of growth? 5 Further 'classic' treatments are in Ranis and Fei. Much basic analysis on the importance of agricultural growth for poverty reduction is based on simplistic premises. For example, it tends to be stated that since the poor are employed in agriculture, agriculture must be the basis of poverty reduction efforts. Against this is one of the basic insights from most data on poverty: that a systematic increase in prosperity tends to be linked to having fewer people dependent on agriculture for their livelihoods. The key question in this respect is - how to get them out of the agricultural sector in a sustainable way?

As it is one of the clearest analyses of this issue, we will build on the work by Eswaran and Kotwal, in a series of academic papers and a monograph, with excellent economic theoretical insights, but with not a single equation in sight. Its relevance lies mainly in the clarity of the questions asked and answered; most of the points made have obviously been made by others as wel, but rarely as crispy and convincing. In this section, we briefly summarize the key points they make, and give a flavour of the analysis. Their analysis is done with India in mind. In the remainder of this section it will then be discussed how applicable these insights are for other parts of the world, most notably, for Africa.

Eswaran and Kotwal analysis can be thought of as a Lewis model within a proper general equilibrium framework. It also drops some of the most difficult assumptions underlying Lewis' original analysis, but that have remained present in many of the subsequent contributions. To put it bluntly, there is no assumption that rural labour markets work in a way, equivalent to agricultural workers spending too much time sitting under trees surplus labour that can be freely extracted. Furthermore, industrial workers are not choosing to eat shirts during a critical phase of the growth process i.e. that when total output in agriculture is going down, they are willing to shift to consuming less food and more manufacturing goods.

Theoretical framework

The EW-model assumes a two-sector economy, industry

and agriculture. There are two goods, shirts and food. Production in both sectors is characterized by different constant returns to scale production technology using labour, as well as using land in agriculture. There are land owners and workers in the rural economy, and workers in the industrial sector. A crucial assumption is made to make preferences lexicographic: people will first need to have enough food before they will buy shirts. It captures an Engel effect that richer people will spend less on basic essentials, but by making it more extreme, its relevance comes out more directly. An alternative means of looking at this is to state that there is no circumstance in which for very poor people, lower prices for manufactured goods cannot induce to them to cut back on essential basic commodities. Even though in reality it may not be as clear-cut, it takes seriously that poverty is related to deprivation in essential food intake. As EW show, relaxing the assumptions does not fundamentally change the result, but makes the dichotomy in implications under different processes of technological and other development marginally less striking. There is also initial inequality in this economy. Some the rich have assets such as land; the poor only have labour. At first, the poor only eat food, since they do not have enough to satisfy their basic needs; once sated with food, they do not eat more. So there is a maximum level of spending on food, and a poor person is someone who only spends on food.

It is further assumed that there are clearing and integrated labour markets across rural and urban areas. It means that people are indifferent between working in agriculture and industry: contrary to Lewis, these markets are not perfectly segmented, but integrated. Clearing product markets mean that demand equals supply in each. All these assumptions on markets imply that poverty will go down if labour demand increases, increasing real wages. In other words, real wage increases for the initially poor determine whether poverty declines. But how does this work? The set-up can be used to contrast a number of alternative strategies to achieve this. Understanding the context and situation in which these different strategies are effective ways of reducing poverty will also help up to understand how important it is to focus on agricultural and rural development. First, consider a closed economy, and the policy considered is (neutral) industrial progress via Total Factor Productivity growth. Under these assumptions, EW show that this implies that more shirts are being produced for the same amount of labour. Prices for shirts will go down, but crucially: the poor do not care for these cheaper shirts, since they still do not have enough food. The result is that there is no incentive for anyone to move out of agriculture, since total food supply would go down, and demand for food would go up. In the end, the TFP growth only benefits the 'rich' that have enough food and are already consuming shirts: they can consume more. The marginal product of labour in industry goes up, but the price of shirts goes down. Employment is unchanged, and nominal wages, food prices and therefore real wages for the poor are the same as before. Poverty is simply unchanged, despite TFP growth in the industrial sector.

Next, consider again a closed economy and a policy of neutral technological progress in agriculture. More food is being produced for the same labour. This is obviously of interest to all the workers there is more food for the same amount of work. Once there is more food consumed, some may cross the threshold and be sated with food, and become interested in buying shirts. The result is that shirt prices are being pushed up. This creates incentives for firms to expand production, and attract more labour to deliver this increased production. Higher demand for labour will require increasing nominal wages.

Rural wages will move up as well, while food prices will go down somewhat since the higher production and the shift to shirts by some previously poor consumers. In equilibrium, labour will have moved out of agriculture into shirt production, and higher equilibrium real wages will imply a reduction in poverty. The contrasting results are striking and lead to the conclusion that in a closed economy, growth in agriculture may well be essential for poverty reduction, while industrial progress has no impact. The presence of demand linkages is the key factor, but for poverty reduction, the relevant linkages are only via commodities consumed by the poor. Mellor and associates have long emphasised this process as well. But there is a difference here the issue is not just growth linkages but also the link with poverty. Agriculture is then the central engine for poverty reducing growth.

The results are nevertheless strongly affected by the assumptions about openness. In an open economy, the central demand and supply constraints do not matter anymore for traded commodities. Basic food staples can typically be imported, while shirts can be exported. Therefore, assume that both goods are tradable goods, so that only world prices matter.

We can now revisit both cases. First, consider the impact of industrial progress. More shirts are being produced for the same labour input, but prices of shirts remain the same as world prices are not affected. Firms have an incentive to expand production, so the demand for labour and nominal wages will increase. Even though food supply goes down, in this case workers can move since food imports can go up. So the marginal product of labour goes up in agriculture as well, allowing rural and urban wages to increase in nominal and real terms. Food imports with higher real wages will mean that more food is being consumed, and some workers will start consuming shirts. The result is that poverty declines. Secondly, the impact of agricultural technological progress is now very similar to industrial progress. The demand linkages are not crucial anymore for the link between real wages and output, and real wages increase with more people buying shirts than before.

To put it simply, poverty reduction can then be achieved by any source of increased domestic competitiveness relative to the rest of the world, a fundamentally different resul to the case of a closed economy. It also creates other vulnerabilities: one's trading partner's productivity growth and any loss of relative competitiveness 'dynamic comparative advantage' could have negative implications in terms of poverty i.e. it has the effects industrial or agricultural decline. Keeping up, or more in general, being able to grow faster in productivity terms than your trading partners is essential. Eswaran and Kotwal, writing near the time of the first moves towards openness in the early 1990s, argue strongly that the evidence is supportive of much potential for India in this respect, especially in growth via technological progress in industry, and that an open trade model could be beneficially assist poverty reduction, even if direct progress in agriculture and the rural sector remained limited. At the same time, technological progress in agriculture has a useful role to play.

Even though the analysis may seem relatively simple, it helps to illustrate some central insights. First, in any scenario, the defining feature of poverty alleviation appears to be linked to the gradual absorption of labour by the non-agricultural sector. There is nothing specific about rural poverty, beyond that it is typically a sign of slow growth. But it goes beyond this: it may also be a sign of the 'wrong' growth forces. Secondly, to understand the role of the agricultural sector in poverty alleviation, it is important to take into account openness. Indeed, the emerging context of globalization changes the dynamic of our understanding of the role of agriculture; it offers more opportunities even though its pitfalls and risks will also need to be properly understood. If the economy is closed, the pattern of growth is crucially important for poverty alleviation. In particular, agricultural growth can then deliver poverty reduction, and given demand linkages, more straightforwardly than industrial progress. Thirdly, introducing more heterogeneity in labour, production technology and sectors does not change the underlying logic see Eswaran and Kotwal. However, if technological progress is not 'neutral' or labour intensive, but instead less labour intensive, then the model would predict a far lower impact on the poverty reduction, since labour demand and real wage increases will be more limited. Eswaran and Kotwal extended the model further with the introduction of a service sector and non-tradable goods the interactions are more subtle but not undermining the main insights. What does this imply for the importance of agricultural growth? In general, it could play a useful role: in many contexts it may be a most suitable mechanism for poverty reduction. However, one should be careful. Technological progress in agriculture can be thought of as involving the discrete adoption of new inputs, such as new varieties, fertilizer or pesticides. It involves some unpleasant arithmetic: much progress in agriculture itself is a 'once and for all' change. For example, fertiliser adoption may well add about 20 percent higher returns to particular crops, but this is not in itself leading to a persistent year-on-year growth in yields. Any growth effect will largely depend on the presence of very substantial linkages. Some have argued that these linkages are very large in developing countries, although this is largely based on evidence on particular countries using particular methodologies that could be and have been criticized.

The available evidence on India is nevertheless illuminating in this respect, without being conclusive. In a number of papers, Datt and Ravallion and Ravallion and Datt discuss the growth and poverty alleviation experience of India, based on state-level averages, using data until 1994, i.e. largely before the gradual opening of the Indian economy, even though overall growth had picked up during the second half of the 1980s. In other words, India could be thought of as a relatively closed economy. A first key finding from this work is that, based on studying the sectoral composition of growth and poverty, agricultural growth appears to have been able to deliver substantial poverty reduction, while manufacturing growth did not do so in itself. Clearly, this is a result directly predicted by the earlier model. Unpacking this evidence showed that yield increases in agriculture had indeed a strong impact on poverty. Secondly, poverty decreases were strongly influenced by particular 'initial conditions'. One way of interpreting is in terms of linkages: a higher level of a particular characteristic, such as the state of technological progress, delivers *ceteris paribus* higher year-on-year growth when its impact works itself through the economy in a self-reinforcing way. The evidence suggests that the key initial conditions were higher health and education presumably correlated with underlying productivity of the workers, and better initial conditions in rural areas including less landlessness and higher initial yields, reflecting better technology. This evidence is clearly consistent with the higher poverty impact from factors increasing the productivity of workers in general, but with a clear role for channelling these growth factors through the rural sector. This is consistent to what was predicted in a closed economy, without necessarily proving that the model is a fully correct representation of the situation.

The opening of the Indian economy and its reasonable success in terms of engaging in the world economy has removed the fundamental demand linkage constraint suggested in the EW model. The sectoral location of productivity growth is becoming less relevant for poverty reduction, provided it is labour-intensive or neutral. The substantial investment in technological progress in other sectors in recent years has opened opportunities for poverty reduction, even though it is likely that the main poverty reducing effects from some of these high-tech service industry investments is likely to have to come from growth linkages towards more low-skill sectors such as construction.

Relevance for Africa

Are these results relevant for sub-Saharan Africa? In the Eswaran-Kotwal model, if growth is driven by agricultural growth that is technologically neutral or labour-intensive, then the scope for poverty reduction will be larger. Land is typically not highly unequally distributed in most African countries, with relatively low landlessness and in some countries, such as in Ethiopia, there is a remarkably equal land distrubution. Productivity increases will reward the poorer farmers as well, so that growth may have substantial poverty impacts. The changing circumstances in African agriculture, in terms of allowing many factor and product markets to work more freely, especially in terms of removing much of the urban bias, has improved the opportunities for this process to materialize. A 'green revolution for Africa' may have then substantial returns.

However, the *necessity* of agricultural growth to deliver both growth and poverty reduction is not as clear-cut. As in the model, opening up the economy has removed the crucial dependence on progress in agriculture to deliver poverty reduction, as the crucial demand linkage with agriculture is removed: growth in other sectors, provided it is labour-intensive, can similarly promote poverty reduction. To assess the case for the crucial role of agricultural growth, the growth opportunities for Africa have to be considered.

Recent work by Ndulu et al. provides the foundation for a suggestive three-way description of the growth opportunities of sub-African countries; a similar description is used in Collier. They distinguish, in terms of growth opportunities: first, resourcerich economies, secondly, coastal and other well-located countries and finally landlocked economies without natural resources. Each of these groups have very different problems at their core, when trying to boost growth and to reduce poverty. For resource-rich economies, the key issue is to manage their wealth: how to translate the underlying wealth controlled by the nation into the basis for sustainable and shared prosperity. The key problems they tend to face are Dutch disease and governance problems and are more likely to be ravaged by violent conflict think Nigeria, Angola or Congo.

For coastal and other well-located countries, the challenge is very different. They have no natural resources, so no immediate source of wealth. It needs to be generated. They have two production factors they can put to good use: they have people and they have their location to their advantage. Much of coastal Africa, not least Ghana, Cote d'Ivoire, Kenya, South or Africa, would spring to mind. Their main challenge is how to take advantage of the opportunities offered by their location. They are countries that in principle should be able to take advantage of world trade opportunities, so their priorities are likely to have to include building up trade infrastructure, managing market institutions and regulation, investing in skills and support the formation of well-working labour markets. These are very different challenges, but globalization offers serious opportunities for them. Without working on their constraints, they are bound to be left behind, but the potential is there.

This leaves the land-locked, resource poor economies without natural resources. They are suffering most from the agglomeration effects: they have little to offer, and they totally depend on their neighbours to overcome these effects. Matters are made worse if their better located or better endowed neighbours are mishandling their economies, or indeed if they are in conflict with these states. All these factors are creating further negative externalities. Examples are Burundi, Burkina Faso or Ethiopia. So when is agricultural growth essential? How does agricultural growth then fit in this framework? First, take the resource-rich countries. Agriculture is unlikely to be an essential source of growth. Nevertheless, such an economy needs to find ways of diversifying and building up their productive capacity, and agriculture could play a role in this. In this context, the burden of agricultural growth to drive overall growth is not present, so that efforts for intensification or diversification can have a much more 'propoor' bias. This could involve a focus on smallholder agriculture, for example by supporting new technologies and activities with higher labour productivity. But clearly, there are a variety of ways to encourage the distribution of the wealth of such a country, and it would be hard to argue that stimulating agricultural growth is essential. Moreover, investing in rural areas, including in basic services such as health, education and infrastructure, could be an effective alternative form of redistribution, and may have higher long-term returns in terms of transforming the economy than a narrow focus on agriculture. For example, there would be less burden on ensuring that these investments are largely in high potential areas.

Secondly, consider the well-located economies, who are best placed to take advantage of world economic opportunities. Managing their comparative advantage, via labour markets, skills, regulation and investment climate, is most essential. The role of agriculture is similar to the EW model in an open economy: 'industrial' progress is most likely the best route, and a vehicle to take advantage of trade opportunities. The role of agriculture is then more subsidiary: it makes sense to encourage progress in agriculture as well, if only as a means of managing the pull out of agriculture when trade-based growth takes off. Skill creation via better health and education, also in the rural sector, is then most helpful as well, if only since it will facilitate the development of better skilled labour that can in due course be absorbed. The experience of Indonesia in the late 1970s and 1980s is rather reminiscent of this, with active rural policies, but ultimately leading to more absorption in the urban economy. For African economies, the key challenge is to overcome their relative marginalisation in the world economy, as late-comers are put to a disadvantage in competing with countries that have already an established industrial base, such as many Asian economies. It may require specific support for manufacturing industrial development for this potential to materialize but it is still the best route for growth.

The landlocked, resource-poor countries are a rather different problem. In many cases think of Ethiopia or Burkina Faso the agricultural base is at best highly vulnerable. But their risk of total marginalization relative to the world economy is also highest. They are mainly dependent on the ability of their better located neighbours to pull them into tradeoriented opportunities, often involving migrant labour. In terms of active policies, the opportunities are limited: infrastructure and skill creation are sensible, but as locations for investment are likely to remain down the pecking order for a long time, not least with many better located neighbours still only barely integrating with the world economy at present. As a consequence, the best way to think of these economies is as if they are effectively closed economies, irrespective of active trade liberalization. As the model predicted, agricultural growth is then essential for both growth and poverty reduction but don't expect any miracles. Technological progress in agriculture has to be actively pursued, as well as other measures to raise rural productivities as the main way of delivering growth that also has clear poverty reducing impacts. Rural and agricultural development is likely to be hardest here, as they are often agriculturally more vulnerable areas, but it is here where it has to be attempted with most vigour. In order to achieve poverty reduction, there are also likely to be important trade-offs between stimulating agriculture in the high potential areas as likely to be required for growth with promoting it in more marginal areas, where poverty may be highest.

The contrast between, say, Ethiopia, and India is striking here. Ethiopia may have been trying to open up in recent years, but with neighbours that are better located but not taking advantage of trade opportunities and in general, with whom relations are at best frosty such as in Somalia or Eritrea. This limits its options dramatically. At best it can create basic infrastructure, create skills but growth is likely to have to come via agriculture to encourage any systematic and persistent decline in rural poverty. Its population cannot take advantage of growth opportunities in neighbouring countries to trade or even migrate. India has surely pockets and even states that face similar local natural resource constraints and other problems to Ethiopia. But its economy is broadly integrated and with some states taking advantage of its increased openness and location, growth externalities and employment opportunities provide options for these lagging states to at least take partial advantage of the overall change, with some beneficial impacts on local poverty reduction.

Market Failures and Poverty Traps

The previous section has focused on a macro or general equilibrium narrative on the role of agricultural and rural development in growth and poverty reduction. Changing contexts, not least the increased openness of developing countries and the move towards more market-based economies changes the role of agriculture and rural development in growth and poverty reduction, not least because urban-rural and sectoral interactions have to be properly integrated in the analysis. However, this preceding analysis is conducted using the assumption of well-functioning factor and product markets. As much of the rural development research in recent decades has highlighted, market failures are prevalent, even despite the removal of many policy-induced market imperfections, such as in agricultural product markets. Resources are then not allocated efficiently, until marginal returns to different factors of production are equalised. Factor markets, including for labour, serve different people differently, resulting in heterogeneity in the extent to which people with different initial circumstances can take advantage of opportunities, including those offered by growth. Rural non-market institutions may have developed to partially substitute or compensate for market failures, but not completely.

This has important implications, not just for the extent of growth that can be achieved, but most importantly for our purposes, the extent the poor can partake in growth. While the underlying taxonomy of general growth opportunities in different parts of sub- Saharan Africa developed in the previous section is unlikely to be fundamentally affected, the ability of these growth narratives to deliver poverty reduction and broad inclusion in the growth process definitely is.

In this section, we try to put some of the key lessons from the extensive micro-level research on rural households and institutions from the last few decades in the broader context of poverty reduction and growth. The key question is: what do we know about the key constraints of the rural poor to participate in or contribute to growth? Much of the recent academic literature on rural issues has explored many of these market failures in factor markets, such as land, labour, credit and insurance. This typically forms the core of much of the teaching offered in graduate schools in the microeconomics of development.10 However, the key issue for our purposes is to explore how micro-level issues can be put into the bigger picture of growth and poverty reduction. In the remainder, we will consider the question of what would cause some poor people to remain locked in rural poverty, even if countries that manage to start growth processes, intensive in terms of labour. Furthermore, for countries that are absolutely dependent on rural development efforts for poverty reduction via growth, such as land-locked resource poor countries with 'poor' neighbours, the need to unlock rural potential means that we should especially carefully examine what may cause growth in particular rural areas to stay behind. We will focus on three instances that illustrate a more general principle and finding in recent theoretical work, with much suggestive empirical evidence "initial poverty and market failures conspire to keep some poor people persistently poor or even in a poverty trap." I will focus on three problems linked to market failures which may induce these processes: access to capital credit market failure risk insurance market failure and spatial externalities curse of geography.

Credit market failure and poverty traps

The most obviously observable market failure is the failure of credit markets to conform to the assumptions of perfectly competitive markets. Under perfect and complete markets, anyone with a profitable project should be able to get a loan at the current interest rate. If markets were perfect and efficient, no bank would ask for collateral to secure the loan. In practice, without collateral, one typically would not get the loan. Collateral requirements can be understood as an important means by which credit markets handle the central problems that bedevil these markets: asymmetric information, such as moral hazard and adverse selection, and enforcement problems. Since imperfect information means that borrowers may not be able to know which projects are more risky among many risky projects, or whether lenders will implement other actions than initially committed to after the loans have been granted, collateral may be asked for to secure the loans.

Collateral may also help to enforce the repayment of loans.

Starting from initial asset poverty for some, it is obvious that this may be a market failure that is particularly hurtful for the poor, by excluding them from profitable opportunities.

A number of careful, suggestive models, most notably Eswaran and Kotwal, show its key implications: the rich do not just earn more income because they have more assets, but also they can use assets more efficiently. Market failures force some of the poor to be inefficient, and they exacerbate any initial inequality some of the poor may stay behind. 11 In this discussion, we use the issue of collateral as a heuristic device to show the differential impact on the poor of credit market failures, without arguing that this is necessarily the key failure. Most other failures in credit markets can be shown to result in a specific disadvantage for poorer households. A useful discussion of different market failures in rural credit markets and what to do about it is in Besley or in Ray.

There is much suggestive evidence that similar processes are common in agricultural settings, often linked to credit market failures. A key prediction of this model is that the marginal return to bringing more land into production by the poor outweighs that of the rich, and that average output per hectare is larger for the poor than for the rich. This negative correlation between cultivated land area and output per hectare is commonly observed in developing countries. Binswanger et al. provide a comprehensive overview of the evidence and look into different explanations. Land quality heterogeneity is certainly part of the story, but factor market failures, including related to credit, are likely to be relevant as well.

The model described is effectively a static model – but its potential dynamic implications are intuitively appealing. Starting from some inequality in assets, those with more wealth earn higher returns and plausibly can accumulate at a high rate while the poor enter into technology or activity portfolios with lower returns, and may not be able to start accumulating any wealth. This intuition is at the basis of a number of growth models leading to poverty traps for some and accumulation for others. Banerjee and Newman show the adverse impact of asset inequality on growth, linked to credit market failures. When threshold levels of assets are needed to enter into different activities, then entry into profitable activities is closed off for those with limited assets, and they are trapped in poverty, while others can climb the occupational ladder. A poverty trap is an equilibrium outcome and a situation from which one cannot emerge without outside help, for example, via a positive windfall to this group, such as by redistribution or aid, or via a fundamental change in the functioning of markets. Much other work suggests poverty traps and overall efficiency and growth losses due to poverty and inequality combined with credit market failure, where some people are unable to exploit growth-promoting opportunities for investment. This credit market failure model is also a central part of the narrative underlying the 2006 World Development Report it discusses also some suggestive evidence of its implications.

If credit market induced poverty traps are present in rural areas, or more in general, underinvestment is present, limiting growth. Furthermore, if growth picks up but if access to profitable opportunities requires some minimal investment e.g. being able to invest in the migration of some household members or sunk costs for a newly profitable activities, then credit market failures may result in the exclusion of some of the poor from the benefits from growth. Proving the existence of credit-induced poverty traps is difficult, but there is definitely evidence in rural Africa of entry costs in particular activities and assets in the presence of limited access to investment capital, leading to poorer households holding less profitable portfolios.

What to do about them is less clear. Credit market interventions have long been a favoured intervention, in recent times mainly via microfinance schemes, although other interventions may well be better and more useful to resolve credit market failures, see Besley. It is also not altogether clear that the poorest are benefiting most from microfinance and more needs to be learned from careful evaluations of specific products that could have the highest impact. For Africa, the key question is also whether microcredit is not overrated as a means to fostering inclusion of the rural poor in the economy: often, if applied to rural settings, it is seen as a means of helping to get the rural poor into off-farm business activities, usually with restrictions on the eligible activities via the support offered by the microfinance institution. Even though credit market restrictions may exclude them from certain profitable opportunities, it is unlikely to be the case that large scale poverty reduction is going to be achieved by making more and more people dependent on enterpreneurial activities. In most economies, the transformation towards a higher income and low poverty economy has been achieved by an increase in wageemployment, via higher labour demand, with farmers become employees in industry and the service sector. It is likely that in the long-run, the highest returns are to be obtained from health, education and skills, not least in Africa economies that are resource-rich or whose potential is in manufacturing exports. Of course, the transition is likely to be helped by allowing some people to take advantage of enterpreneurial opportunities in agriculture and the nonfarm sector, but it is unlikely to be a large-scale succesful route.

There is a dilemma here as well: as long as growth is not taking off, the poor are likely to be helped more by encouraging enterpreneurial activities, as labour demand is not picking up. In most African contexts, returns to education are convex: low for primary and possibly high for higher levels of education, so that investing in education under circumstances of the 1980s and 1990s, with low wagelabour demand, is unlikely to offer rapid routes out of poverty, but if growth were to pick up in a sustained way, this could change. Failing that, microcredit schemes could offer a solution for many poor, but without growth it is not offering real scope for large-scale poverty reduction. In any case, and at best, microcredit schemes with more flexibility, that do not try to tie people to particular enterpreneurial activities, but that respond to the general financing needs of families, may be more effective.

Insurance market failures and risk induced poverty traps

Another serious market failure impacting disproportionately on the poor is the lack of insurance and protection of the poor in the face of risk. Complete insurance markets or, to be technically more accurate, complete state-contingent markets is another assumption for perfect markets that tends to be violated in practice. Problems with asymmetric information and enforcement issues, not dissimilar to those causing credit market failures, are again typically responsible for the limited spread of insurance mechanisms in developing countries. Even if they wanted to, the poor could not get any insurance for most of the risks they face.

Uninsured risk causes considerable hardship to the poor. Developing countries are still characterised by a high incidence of natural disasters, drought, conflict and insecurity, as well as economic shocks, such as commodity price shocks and currency shocks. Health problems are widespread, as are pests in agriculture. It is commonplace to view these problems as 'transitory' problems, requiring temporary solution, such as some form of safety net, after which one should get back to the bigger issues of development. For the policy maker it also often means that it is just a social issue that should not distract the key macroeconomic policy makers from the bigger issue of how to stimulate growth in the economy and widespread poverty. However, this is misleading. There is increasing evidence that risk and shocks are a cause of lower growth, especially resulting in lower growth of incomes of the poor and possibly poverty traps. Focusing attention on the poor could then again be contributing to both growth and equity; in any case, it could be instrumental to ensuring that the poor can benefit from emerging growth.

Households in developing countries have developed sophisticated mechanisms to cope with risk. Typically, one could consider two types of responses: risk-management strategies and risk-coping strategies. Risk management strategies involve trying to shape the risks they face by entering into activity portfolios that are more favourable in terms of risks. For example, entering into low risk activities or diversifying into portfolios of activities with differing risk profiles growing more drought resistant crops, entering in petty trading or firewood collection, seasonal migration, etc. Risk coping strategies involve activities to cope with consequences of risk in income. Two types are commonly observed: self-insurance using savings, often in the form of cattle or small ruminants, to be sold off when the need arises, and informal mutual support mechanisms, where members of group or community provide transfers to each other in times of need of one of its members, typically on a reciprocal basis. These strategies are not without cost: income risk management strategies result in a reduction in mean income to have face risk and variability in income, while adjusting asset portfolios to cope with risk typically involves investing in liquid assets with lower returns rather than in productive illiquid investment. This affects their long-term income and their ability to move out of poverty. Indeed, there is growing evidence that these strategies imply substantial efficiency loss for the poor, which the rich typically better protected via insurance, asset and credit - do not have to endure. Morduch documents how more profitable technologies are not adopted because they are too risky in a particular setting in India. The same farmers have been found to hold livestock as a precaution against risk even when more productive investment opportunities exist. Rosenzweig and Binswanger found that the loss in efficiency between the richest and poorest quintiles in their sample from India was more than 25 percent, attributable to portfolio adjustments in assets and activities due to risk exposure. In Ethiopia, Dercon and Christiaensen have found that modern input adoption is lowered due to the downside risk related to input loans, whose repayment is strictly enforced, even when rains and harvests fail. Over time, this results in substantial efficiency losses, affecting the poor disproportionately.

These risk management strategies may trap the poor in poverty: to avoid further destitution, they are forced to forgo profitable but risky opportunities, and with it the opportunity to move out of poverty. Even so, they cannot fully protect themselves: there is much evidence that although the strategies contribute to less variability in consumption and nutritional levels, they are still not able to cope with some serious, repeated shocks, not least those affecting whole communities, regions or countries. These uninsured shocks typically wipe out assets, pushing the households down the asset distribution. They could be pushed below some critical threshold, trapping them into poverty from then on, for example due to the risk strategies they then need to follow to avoid further destitution, or due to other processes. There is growing evidence that these processes are an important cause of poverty persistence and possibly permanent traps in developing countries. Jalan and Ravallion investigated the presence of poverty traps using data from China, and although they did not find a pure poverty trap, they found that households took several years to recover from a single income shocks, and that the recovery was much slower for the poor. There is also related evidence from Africa. Dercon using panel data from rural form of savings in rural Ethiopia, to recover to the levels seen before the 1984-85 famine. In a careful study, Elbers et al., use simulation based econometric methods to calibrate a growth model that explicitly accounts for risk and risk responses, applied to panel data from rural Zimbabwe. They found that risk substantially reduces growth, reducing the capital stock (in the steady state) by more than 40 percent. Two-thirds of this loss is due to ex-ante strategies by which households try to minimize the impact of risk. Barrett has suggestive evidence on poverty traps by looking at livestock holdings of pastoralists in Kenya.

There is also increasing evidence on the long-term implications of uninsured shocks, focusing on health and education. For example, the permanent impact of drought on children is well documented lower adult height, poor education outcomes and therefore lower lifetime earnings. For example, the impact of drought and war in rural Zimbabwe in the early 1980s on a particularly vulnerable cohort of children was estimated at 7-12 percent of lifetime earnings or more.

All this evidence points to the important consequences of lack of insurance and protection in rural settings in developing countries, particularly affecting the poor. Given that the root cause is again a market failure, exacerbated by poverty, there is a clear case for interventions that are potentially both poverty reducing as stimulating efficiency and growth; in any case, they could ensure that the poor can more effectively take part in growth processes. In developed countries, not least in Europe, the failings of the insurance markets are largely resolved via some form of universal social insurance and substantial direct means-tested transfers. For developing countries, this is not likely to be cost-effective, involving high administrative costs and high informational requirements.

To put it simply, the means for such systems are unlikely to be available. Many responses can be considered, from reducing the risks faced by rural households e.g. by preventative health services, or better water management for agriculture, strenthening of existing responses (such as investing in better savings product provision, or better functioning assets markets, such as for live animals) and improved forms of insurance and broader social protection in the form of safety nets. While each have their problems and advantages, in recent years a number of specifically interesting initiatives have been taken with respect to insurance, even if their benefits are not quite substantiated yet.

Spatial effects

Another common cause for market failure is the presence of spatial externalities. Externalities are said to be present if economic or other interactions create social gains or costs beyond those taken into account by those involved in the interaction. The standard example is environmental damage from production involving pollution not accounted for by the buyers and sellers of the commodity produced. A more general phenomenon in developing countries, that can be best understood in terms of externalities, involves geographically defined areas that appear to stay behind poor neighbourhoods, or even poor regions or countries. If one looks at the performance of the developing world, it has been striking over recent decades that some developing countries largely in Africa appear to have become increasingly marginalised, with low economic growth, persistent population growth and generally, persistent poverty. Less studied but at least as important is that even in countries were growth is high, there appear to be areas that systematically stay behind and do not benefit from the overall economic growth in terms of income growth and poverty reduction. Certain regions in China and India may well fit this bill.

Much less documented but no less true, the geographical disparity in growth and poverty reduction performance within African countries is similarly present.

Such disparity may well be explained in terms of theories emphasising agglomeration or location effects, predicting that firms will exploit increasing returns resulting from the presence of externalities to locating in the same geographical areas, implying that firms would locate in clusters. The corollary is that some less attractive locations may have missed the boat: not only would they not get the required investment, any capital present may well move out to capture the higher returns elsewhere. For those areas that missed the boat, there is a negative externality from the success of other areas.

Clearly, this is a form of a poverty trap: although initially these areas may not have been very different, once they missed the boat, they can only escape by a serious exogenous shock or massive effort. They face a substantial threshold that they need to overcome to attract or retain capital for accumulation.

Other explanations similarly emphasize externalities related to the specific local context, for example low local endowments in terms of public goods, common property resources and private asset holdings. If growth processes require a certain threshold of local endowments to take off, then poorly endowed areas may well find it hard to escape poverty. Proving this is difficult. There is some evidence for China and a few other countries. For Africa, the role and effects of remoteness on growth and poverty in Africa are well discussed in Christiaensen et al. with evidence from a number of countries.

Recall that these externalities are again market failures that specifically affect those at the lower end of the asset distribution this time with assets broadly defined to include public and environmental goods. Given that poverty traps are identified, this empirical evidence would justify 'poor areas' programmes massive investment programmes in particular deprived areas to build up locational and community capital. However, these empirical studies lack sufficient detail and a clear narrative about how these externalities come about. More evidence would be needed to guide and prioritize the type of interventions that would be most beneficial.

15 Jalan and Ravallion identify geographic poverty traps in rural China during the 1980s, finding that community characteristics affect the income growth performance of otherwise identical individuals, controlling for latent heterogeneity. There results showed that in some areas living standards were falling while elsewhere *otherwise identical households* were enjoying rising living standards, an effect entirely due to externalities from the initial community characteristics.

For example, most rural 'poor areas' typically are characterised by remoteness often linked to the lack of roads and communications infrastructure. One of the most common donor policy responses is to build roads into poor areas. While undoubtedly bringing some benefits to remote communities, it is not necessarily the case that this is what is needed to unlock the growth potential of an area. In some countries, there is evidence that this may well be an appropriate response. Still, historically much road building in developing countries has been *in response* to local economic growth or at least in recognition of some growth potential such as cash crops or mining, and it was not the main cause of growth. Alternatives, such as irrigation, health or educational schemes may be more important to unlock their potential.

In any case, just doing a little is not going to help in order for such areas to catch up, substantial levels of investment would be needed to lift them over the threshold. It may well be the case that creating opportunities to leave may be a superior policy. Migration nevertheless also involves costs, so this may be difficult too. If these thresholds are substantial, then growth opportunities elsewhere may simply bypass many rural poor. But it is nevertheless the prime example of a potential rural poverty trap for which solutions have to be firmly considered in the context of rural-urban and other linkages, and a narrow focus on rural areas may be ineffective.

Conclusions

Contrary to most of the rest of the developing world, poverty levels have remained stagnant in sub-Saharan Africa in the last few decades, in a context of slow growth. As most of the poor are living in rural areas, it may lead to the conclusion that agricultural growth and rural development policies have to be at the core of growth and poverty reduction policies.

In this paper, we used a 'macro' and micro-perspective to discuss some of the key concerns and issues involved in an African context. Using a framework based on Eswaran and Kotwal, and evidence on growth opportunities in Africa based on Ndulu et al., we concluded that agricultural growth is likely to be essential for landlocked and resource-poor economies in Africa, even though growth opportunities via agriculture are likely to be difficult. In others, in particular those with good location for engagement into manufacturing exports, or resource-rich economies, agriculture is not the crucial constraint. Even if rural development policies are likely to be crucial to allow the gradual transformation of the economy, the pressure for agricultural growth to be an engine of growth is not present.

This analysis is however based on well-functioning factor and product markets, allowing specifically the poor to take advantage of opportunities irrespective of their endowments.

Perfect markets in rural settings are not the appropriate assumption. There is considerable evidence that this implies that poor may remain excluded of profitable opportunities to grow out of poverty, even if growth has picked up in the economy, potentially leading to poverty traps. Rural development policies targeted on the poor, and possible including stimulating agricultural production by poor farmers may then be part of effort to make growth more inclusive. The mechanisms by which this exclusion may happen are well understood, and we discussed credit and insurance market failures, as well as spatial effects, in the form of 'poor areas'. The evidence on the appropriate responses is still only emerging, and clearly more experimentation and research is needed.

Chapter 7

Linkage Between Rural Poverty and Agricultural Productivity

Improvement in agricultural productivity can reduce rural poverty in a major way. In India, various studies inferred that rural poverty is inversely related to the agricultural productivity and argued that agricultural productivity growth reduces rural poverty effectively as it generates income for poor farmers. Indian states are characterized by high variations in rural poverty levels, particularly, in Uttar Pradesh which is the most populous state of India. The economy of the state is rural-based, predominantly agrarian, with 80% of population living in villages. The incidence of rural poverty in the state is very high as compared to most of the other states of India. It demands for an in-depth analysis of rural poverty in Uttar Pradesh and the role of agricultural productivity in reducing rural poverty. Present study attempts to estimate rural poverty and disparity in it within the state. The linkage between rural poverty and agricultural productivity has also been examined. The study analyses that there exists no significant differences in the incidence, depth and severity of rural poverty across the regions, while variations are significant across the districts. The linkage between agricultural productivity and rural poverty are found to be significant in the state. This shows that as agricultural productivity increases, the incidence, depth and severity of rural poverty decreases in all the districts.

The findings of regression analysis further provide evidences that the negative impact of agricultural productivity on the incidence, depth and severity of rural poverty is statistically significant across the districts of Uttar Pradesh.

Improvement in agricultural productivity can reduce rural poverty in a major way. Many researchers have studied this aspect of the overall socio-economic development of different countries and the states in these countries involving India. They, however, could not arrive at any specific conclusion. Their findings widely vary in respect of the causes behind the continuation of poverty in the states and regions within. However, quite a good number of studies have pointed out that there is a close relation between agricultural productivity and poverty in rural areas as around 75 per cent of the poor work and live in rural areas, and projections suggest that over 60 per cent will continue to do so in 2025.

Therefore, there is a need to focus on reducing rural poverty and to redirect attention and expenditure for agricultural development that generates employment in rural areas, as agriculture comprises largest component of rural economy in developing countries, particularly in India.

Agricultural productivity growth may have more immediate multiplier effects in improving the well-being of the poor. Wichmann, using a dynamic general equilibrium model showed that an increase in agricultural productivity can lead to a significant increase in the household consumption of the poor. Strong agricultural growth, particularly increased productivity, has been an important feature of countries that have successfully reduced poverty, particularly rural poverty. Gallup et al. in a cross country examination established that one per cent increase in agricultural GDP led to 1.61% increase in income of the poorest quintile, while the corresponding values for the manufacturing and services sectors were only 1.16 per cent and 0.79 per cent respectively. Thirtle et al. concluded from a major cross country analysis that, on an average, every 1 per cent increase in agricultural yields or land productivity reduced 0.83 per cent of people living on less than US$1 per day.

A host of studies shows that higher agricultural productivity in Asia consistently raised farmer's incomes despite declining market prices resulting from increased output. Small and medium sized farmers have not been excluded from these benefits. Similarly, in a developing country like India, agricultural productivity has been identified as an important contributing factor in attaining food security and in reducing poverty. Ahluwalia in a study for India spanning the period 1956 - 57 to 1973 - 74, inferred that rural poverty is inversely related to the agricultural productivity. Mellor also argues that agricultural productivity growth reduces poverty effectively as it generates income for poor farmers. Increased agricultural productivity also creates employment opportunities in farms, although not necessarily resulting in higher wages. Recent studies of poverty reduction in India support the view that poverty reduction is the result of growth within agriculture in both rural and urban areas.

Despite this fact, poverty still continues to be a major concern in India. The magnitude of rural poverty is larger than that of urban poverty in the country. The incidence of rural poverty in Uttar Pradesh is very high as compared to most of the other states of India. UP alone holds 473 million rural poor below the poverty line which is the highest number of absolute poor in the country for the 61st round of NSS. In 2004 - 05 rural poverty in UP is 33.4 per cent, which is higher than all India figure of 28.3 per cent. UP being the most populous state of India comprises about 17.9 per cent of India's rural population and 20.3 per cent of country's poor. The economy of the state is rural-based with 79.2 per cent of population living in villages. According to 1983 and 1993 - 94 rounds of NSS, nearly two-third of rural households employed as agriculture laborers were below the poverty line as compared to half of the households working as casual laborers in rural non farm sector.

In India inter-state and intra-state, disparities are studied in different dimensions but disparities in poverty levels have been examined extensively. Similarly, UP is characterized by large variations in rural poverty levels across the regions and districts.

Bundelkhand region is marked by highest incidence of poverty followed by Eastern and Central regions. Western region has the lowest incidence of poverty and is also the most developed region in the state. Here, the yield of food grain crops is 25.55 q/ha for the Western region, while it is only 19.66 q/ha in the Eastern region.

Thus, an apparent link between the productivity of crops and the income of the people, and hence the poverty, particularly the rural poverty, is observed in UP. The variation in the levels of rural poverty also appears to be linked to the variations in agricultural productivity. Theoretically also a relation appears to exist between these two since in many states and their districts a large number of people earn their livelihood from agriculture, even the productivity of workers, working in factories, industries, trade and commerce depend upon the availability of food. Many workers who earn a bare minimum just to maintain their families can hardly improve their productivity if food production is not sufficiently high and the price which depends on the demand for and supply of food is well within their reach. This apparent link needs to be meticulously looked into before any conclusion on the link between agricultural productivity and the rural poverty in particular and poverty in general is drawn.

Objectives

In the present study we make an attempt to estimate rural poverty and its variations across the districts in UP. The study also attempts to establish the linkage between rural poverty and agricultural productivity in UP. If any such link is found between them the degree of dependence or effect of agricultural productivity on rural poverty is examined. In other words, study also examines how much of the total variation in rural poverty is accounted for by the variation in agricultural productivity. The objectives in this study are set out in the following ways:

- To measure the incidence, depth and severity of rural poverty across regions and districts in UP.
- To analyze the regional and district level variations in incidence, depth and severity of rural poverty in UP, and
- To examine the linkage between agricultural productivity and incidence, depth and severity of rural poverty across the districts of Uttar Pradesh.

Hypotheses

To fulfill the objectives following hypotheses are formulated related to objectives two and three.

H1: There is no statistically significant difference in the incidence, depth and severity of rural poverty across the regions and districts.

H2: There is no linkage between the agricultural productivity and the incidence, depth and severity of rural poverty in the regions and the districts in the state.

H3: There is no effect of agricultural productivity on the incidence, depth and severity of rural poverty.

The method used to test the above three hypotheses and the method to measure the incidence, depth and severity of rural poverty are described in the following section of methodology.

Methodology

Present study estimates the rural poverty over four quinquennial years for the four regions and 48 districts of UP. The estimation of rural poverty is based on the National Sample Survey Organization data for the 43rd 50th, 55th and 61st rounds of Household Consumption Expenditure Survey reports. The 55th round of NSS is not comparable with other rounds of NSS due to modifications in the estimation and survey techniques of NSSO. However, rural poverty has been estimated for the 55th round also for the regions and districts of UP through adjusted pooled State and Centre data of NSSO. Therefore, the estimates of rural poverty are not comparable with other rounds in the present study. The data considered in this study exclude the Hill region, as it became a new state that is, Uttarakhand in 2001. The data for the newly created districts have been combined with the respective parent districts for comparability purposes.

The study takes the help of rural poverty line of UP as defined by the Planning Commission of India. The poor are defined here as those whose monthly per capita income is less than the poverty line as estimated by the Planning Commission of India. The incidence, depth and severity of rural poverty for the regions and the districts are the weighted estimates. The weights attached to the incidence, depth and severity of rural poverty are the multipliers, as suggested by the NSSO and the Planning Commission of India.

Present study uses the widely used Foster-Greer-Thorbecke measure of poverty. This approach assumes that individuals and households are poor if their income or consumption falls below a certain minimum threshold that is, the poverty line. Here, for estimation of incidence, depth and severity of rural poverty, the monthly per capita expenditure data are used, as consumption data are readily available, stable and reliable than the income data especially in the rural sector. The method is also used by NSSO and other researchers.

Poverty Gap captures the acuteness of poverty since it measures the total shortfall of the poor from the poverty line. Individuals above the poverty line have a zero poverty gap. But this does not consider the importance of the number of people who are below the poverty line. That is why it is necessary to use together both the measures of poverty that is, incidence and depth to evaluate the extent of poverty. However, neither of the two is sensitive to the destitute. If a poor individual receives an income transfer from much poorer one, then neither incidence nor depth will change. Therefore, higher order such as severity of poverty is used.

The severity of rural poverty gives greater weightage to those who are further away from the poverty line. The regional and district level variations in incidence, depth and severity of rural poverty have been analyzed through mean variance test, as defined by the ratio of Between-column Variance and Within-column Variance.

Poverty, whether measured by incidence or depth or severity, is related to the level of income of the households. There is an inverse relationship between these two that is, poverty is an inverse function of income. For the rural people, however, the income data are not available directly from any records since most of them do not have any such sources of income for which records are kept. The major source of their income is agriculture which constitutes about 70 per cent to 90 per cent of the income of the rural people.

However, data related to agricultural income of the rural people are not available from secondary sources. Given this limitation, agricultural productivity has been considered here as a proxy for the agricultural income of the rural people. Higher the agricultural productivity, higher will be the income of the rural households, given that the prices of agricultural products do not fall sharply. This is possible because of the implementation of the minimum support price policy by the central government.

Therefore, the present study uses agricultural productivity of 48 districts and four regions that is, Western, Eastern, Central and Bundelkhand in the state separately as a variable that is inversely related with poverty of the rural people. The linkage between agricultural productivity and incidence, depth and severity of rural poverty are analyzed through the contingency tables for each dimension of the poverty measures. The variables have been grouped into different levels such as low, moderately low, moderately high and high for rural poverty and low, medium and high for agricultural productivity for preparing the contingency tables.

The effect of agricultural productivity on different aspects of rural poverty has been measured by regressing rural poverty on agricultural productivity, assuming linear relationship between these two, and then carrying out statistical significance test on the regression coefficient of agricultural productivity.

However, in the year 2004 - 05, this region experienced the highest incidence of rural poverty among all the regions. However, in general, the incidence, depth and severity of rural poverty show a declining trend over the years in all the regions it is observed that for the years 1987 - 88 and 1993 - 94, the districts falling in the Bundelkhand region namely Lalitpur, Hamirpur, Banda, and Jhansi have the highest incidence of rural poverty. Lalitpur, Jalaun and Banda show higher incidence of rural poverty in 1993 - 94 in comparison to previous year that is, 1987-88.

Gorakhpur, Varanasi, Ghazipur and Ballia have depicted a reduction in the incidence of rural poverty. In contrast, in 2004 - 05, the districts show higher incidence of rural poverty in comparison to that in 1999 - 00. Overall, the districts in the Eastern region present a fluctuating trend of the rural poverty over the years. However, only two districts that is, Jaunpur and Gonda have shown a declining trend in the incidence of rural poverty. The districts of the Central region such as Hardoi, Rae Bareli, Fatehpur, Unnao holds high percentage of rural people below the poverty line. Only Kheeri and Hardoishow a declining trend in the incidence of rural poverty.

Other districts show a fluctuating nature in the incidence over the period. In 1993 - 94, the incidence of rural poverty in Lucknow, Unnao, Rae Bareli, Kanpur Dehat and Nagar and Fatehpur is marked with a rising trend in comparison to that in 1987 - 88. The incidence thereafter follows a declining trend in 1999-00 with exceptions of Sitapur and Kanpur Dehat and Nagar and also in 2004-05 with the only exception of Lucknow. In the Western region of UP the majority of the districts are characterized with lowest incidence of rural poverty for the years under study. Only Moradabad and Aligarh show a decreasing trend in the incidence of rural poverty over the years. The rising trend in the incidence in 1993 - 94 over 1987 - 88 might be explained with the rise of the poverty line in the rural areas by Rs. 100. Such elevation of poverty line by Rs. 100 has caused a rise in the percentage of people below the poverty line. In 1999 - 00, although there is a rise in the poverty line by Rs. 123, majority of the districts show a fall in the incidence of rural poverty due to the change in the survey method of the NSSO for 1999-00 round of household consumption expenditure survey. Overall, a fluctuating trend has been observed in the incidence of rural poverty over the years in Western UP. The depth and severity of rural poverty also show similar results for the districts of Western UP.

The declining trend of rural poverty across the districts might be attributed to the poverty alleviation programmes and other schemes implemented in the rural areas to generate employment and raise the income of the poor households. The schemes such as IRDP, SGSY, and Jawahar Rozgaar Yojana have been effectively implemented in most of the districts of UP. Although the poverty levels in the districts have fallen over the years, in majority of the districts, poverty levels have been very high for the year 1993 - 94. In many studies, this phenomenon has been explained as a result of the introduction of economic reforms in India in 1991 - 92. On the contrary, some researchers such as Tendulkar and Jain Joshi and Little attributed it to bad weather resulting in low rainfall.

Therefore, economic reforms and weather conditions were suspected to cause sudden increase in rural poverty. The regional variations in the incidence, depth and severity of rural poverty, shows that the null hypothesis is accepted. The F values are found to be statistically insignificant for the incidence, depth and severity of rural poverty. It means that there exists no disparity in the levels of rural poverty among the four regions of the state over a period from 1987 - 88 to 2004-05. Here, in the analysis, 1999-00 year has not been considered for comparability purposes. On the contrary, the variations across the districts are found to have significant variations in the state, with F values significant at one per cent level of significance, indicating differences in the incidence, depth and severity of rural poverty across the districts of UP.

The inter-relationship between agricultural productivity and rural poverty has been examined through contingency test. From the contingency tables it is found that the calculated values of –2 for each year are statistically significant and different from zero at one per cent probability level for six degrees of freedom. This indicates that our null hypothesis of no association between agricultural productivity and poverty does not hold true. In other words, this means that a significant inter relationship exists between the agricultural productivity and the rural poverty. Lower the agricultural productivity, higher will be the rural poverty in the state. The districts falling under different categories of poverty such as low.

Similarly, the results of the linkage of agricultural productivity and depth and severity of rural poverty across the districts have been presented in Appendix III and Appendix IV respectively. A comparison of the inter-relationship between these two variables that is, agricultural productivity and levels of rural poverty over the years shows that the strength of this negative relationship has weakened over the subsequent five years. This might be due to the fall in the growth of agricultural productivity in the state over the years . The agricultural productivity presented through an inverted U-shaped curve1 during 1987- 88 to 2004-05. The figure shows that the agricultural productivity has been increasing at a decreasing rate in UP over the years. The negative coefficient value of t2 clearly confirms the above statement.

The findings of the regression analysis show that there exists an inverse relationship between agricultural productivity and the levels of rural poverty with statistically significant negative coefficients of agricultural productivity across the districts of UP over the years. It confirms that the rise in agricultural productivity leads to the decline of incidence, depth and severity of rural poverty in the state. The F values for 1987-88, 1993-94 and 2004 - 05 are significant at one per cent while for 1999-00 it is significant at five per cent only. This may be explained due to the modifications in the survey methods of NSS for this particular round, which makes it incomparable with the other rounds of NSS considered in the present study.

Similarly, the total variation explained by agricultural productivity in the levels of rural poverty is high for the years 1987-88, 1993-94 and 2004-05 in comparison to 1999-00. The impact of agricultural productivity in reducing the incidence of rural poverty is higher as compared to depth and severity of rural poverty in all the selected years, as observed from the coefficient values.

Summary and Conclusion

The study concludes that there exists significant difference in the incidence, depth and severity of rural poverty across the regions and across the 48 districts of UP which have found to be significant at one per cent probability level. As rural poverty is still high across the districts in comparison to national average this needs to be addressed and taken care of.

The linkage between agricultural productivity and rural poverty are found to be significant. The –2 values for rural poverty for the selected years are found significant at the district level. This shows that as agricultural productivity increases, the incidence, depth and severity of rural poverty decreases in all the districts. Further the regression results show that agricultural productivity has statistically significant effect upon the rural poverty.

Thus, all these empirical exercises establish a close link as well as cause and effect relationship between these two variables; rural poverty and agricultural productivity. Hence, to remove poverty in the rural areas of the state, the state government should make a sincere effort to improve the agricultural productivity in the tricts of the state, which have been paid only a lip service till date at both the economic and political level. Neglect of agriculture will hamper the growth of industries, particularly agro-based industries and weaken the macroeconomic fundamentals of the state and the country as a whole.

Chapter 8

Agricultural and Rural Development for Reducing Poverty and Hunger in Asia

The International Food Policy Research Institute was established in 1975. IFPRI is one of 15 agricultural research centers that receives its funding from governments, private foundations, and international and regional organizations, most of which are members of the Consultative Group on International Agricultural Research. "A 2020 Vision for Food, Agriculture, and the Environment" is an initiative of the International Food Policy Research Institute to develop a shared vision and a consensus for action on how to meet future world food needs while reducing poverty and protecting the environment. Through the 2020 Vision Initiative, IFPRI is bringing together divergent schools of thought on these issues, generating research, and identifying recommendations.

The Asian Development Bank aims to improve the welfare of the people in the Asia and Pacific region, particularly the nearly 1.9 billion who live on less than $2 a day. Despite many success stories, the region remains home to two thirds of the world's poor. ADB is a multilateral development finance institution owned by 67 members, 48 from the region and 19 from other parts of the globe. ADB's vision is a region free of poverty. Its mission is to help its developing member countries reduce poverty and improve their quality of life. ADB's main instruments for helping its developing member countries are policy dialogue, loans, equity investments, guarantees, grants, and technical assistance. ADB's annual lending volume is typically about $6 billion, with technical assistance usually totaling about $180 million a year.

During the past three decades, Asia has undergone dynamic growth and structural transformation. The incidence of people living in poverty has fallen from more than 50 percent in the 1970s to 18 percent in 2004—representing a decline in the overall number of poor of about 300 million. The incidence of hunger has also exhibited similar dramatic decreases, declining from more than 30 percent to 16 percent over the same period. The continuation of these trends is likely to create a dramatically transformed Asia; it is expected that by 2015, the region will contribute roughly 42 percent of global GDP.

These achievements in reducing poverty and hunger were due in large part to growth in the agricultural and nonfarm rural sectors through the 1980s. One of the keys to East Asia's economic success was increased agricultural productivity resulting from a liberalization of markets and rapid technological change. Agricultural growth impacts rural poverty reduction not only through increased farm incomes, but also through stimulating the nonfarm economic sector in rural areas and small towns. At the same time, nonfarm income from trading, services, transportation, handicrafts, remittances, and small-scale manufacturinghas become increasingly significant and now makes up 51 percent of income in Asia's rural areas.

However, despite Asia's unprecedented economic growth and poverty reduction during the past three decades, hunger and poverty persist, with more than 600 million people living on less than US$1 a day. By 2015, Asia will still be home to half of the world's poor, and best projections indicate that three-quarters of these poor will live in rural areas.

The key to further poverty reduction in Asia is inclusive and sustainable growth, which at its core means rural development, sustainable agricultural growth, and effective natural resource management in the region. At present, the progress that Asia has achieved in reducing poverty and hunger has been uneven.

- First, most of the reduction in poverty has taken place in China and, to a lesser extent, in Southeast Asia; South Asia has seen more limited success. Though hunger has declined significantly in East Asia and the Pacific, it remains a critical problem in South Asia. Thus, it is important to remember there is not just one "Asia," but several, all needing different policy approaches.
- Second, different income groups have benefited differently from the growth process. Across the continent, the rich are getting richer faster than the poor. Among the poor, those just below the dollar-a-day poverty line have benefited from economic growth more than those far below it, implying that recent growth has bypassed the poorest, especially those living in marginal areas such as remote mountainous regions.
- Third, different sectors have benefited differently. East Asia's growth is now contributing to an expanding income inequality between those living in the cities and those living in the rural areas with inadequate access to technology, infrastructure, markets, credit, insurance, and social capital. This growing gap is not economically or politically sustainable over time.
- Last but not least, certain gender caste identity groups remain largely excluded from the growth process. Discrimination of women in South Asia has been shown to be a major deterrent to reducing hunger and child malnutrition in the region.

Exclusion and discrimination on the grounds of caste and

identity also needs to be addressed in Asia both on the grounds of fairness and on the grounds that it hampers economic growth as those who would be very productive in the marketplace often with assets and education are barred solely on the grounds of their group identity.

In addition to the uneven growth process, increasing pressures on natural resources and environment in Asia are also challenging the sustainability of the economic growth in the region. Most people in Asia remain directly dependent on a productive natural resource base for their livelihoods. Unfortunately, population growth, industrialization and urbanization in Asia's economic growth have put increasing pressure on its natural resources and environment. Urban-biased industrial development and nonlocally managed investments in resource extraction within the context of an expanding global trading system are leading to resource depletion. Asia has become an important contributor, as well as a victim, to many environmental problems including global warming, water and air pollution, ecosystem degradation, and biodiversity losses.

Emerging Trends and Challenges

Asia now finds itself at a crossroads, facing challenges and opportunities which, if properly addressed, can allow the region to continue to make significant reductions in hunger and malnutrition. Some of these challenges and opportunities are identified below:

Food markets and agricultural production

Food markets are changing rapidly in Asia, largely as the result of changes in demand due to population growth and economic growth such as increased demand for fruit, vegetables, dairy, and meat and reduced demand for cereals and changes in the structure of the food retail sector such as the entry of supermarkets. These changes are transforming the way farmers interact with markets and are taking place in the context of declining farm sizes. In addition, the increased demand for many food products combined with biofuel demand, resource scarcity, and climate change has led experts to predict that the real-world food prices of most cereals and meats will rise by 20–40 percent, reversing a long-established downward trend. This will have a serious short-term impact on poor consumers in Asia and elsewhere but may also represent a medium-term opportunity for smallholder farmers who can move up the value chain and enter new markets. Thus, the solution may involve increasing food production through the adoption of biotechnologies and other innovations, and improving infrastructure, and creating broader and deeper markets that allow for quality differentiation and more value-added products.

Agricultural energy

Though an increasing demand for agricultural energy has been pushing food prices higher, biofuel production also has the potential to create opportunities for Asian farmers by increasing the demand for agricultural products including biomass products and by providing some farmers with a way to use crop residues and marginal land. Furthermore, because biofuel production can be labor intensive, it may generate additional employment in rural areas.

Smallholder farmers

Enabling asset-poor farmers to produce for and connect to these new markets is a challenge for both the public and private sectors in the region. In the context of declining farm size, important considerations include ensuring that new crops and new technologies are scale-neutral, training farmers in new crops and markets, improving access to credit for farmers without collateral, and investing in supportive infrastructure such as cold storage facilities and the regulation of transport biofuel markets.

It is also essential to design technologies that the poor need and want, and to that end, investing in rice technologies for rain-fed systems where the poor are concentrated is crucial.

Investing in technologies for irrigated rice systems is also crucial in providing a source of low-cost staple crops in the region.

Nonfarm activities

The growth of the nonfarm sector in Asia has been fundamental in allowing poverty to decline amid high population pressure and closed land frontiers. It has also led to substantial reduction in poverty in areas with low agricultural potential. Nonetheless, there is a need to better understand the nonfarm sector so as to more efficiently focus policy initiatives in this area: nonfarm activities are often small, family-owned businesses characterized by low levels of human capital, limited credit, infrastructural constraints, and limited ability to use and enforce contracts. Improved access to credit and policies that provide reliable affordable electricity and telecommunications services would support these businesses. Reforms that encourage the formalization of these businesses are also important in enabling this sector to play the role it needs to in markets that are increasingly quality driven.

Trade liberalization

Asia's agricultural and rural sector has recently been adjusting to the external forces of competition unleashed by agricultural trade liberalization. Globalization offers both a challenge and a chance. For food exporting countries, export trade liberalization will help smallholder farmers increase income and expand production, but a rise in food prices in these countries might hurt poor consumers. The speed of trade liberalization and domestic market reforms should be gradual and subject to close monitoring and short-term adjustments. It is important to develop a strategy to help poor farmers transition to being effective in liberalized economies. Building the assets of poor farmers has to be the first step, creating new market opportunities and improving their competitiveness is the second, and regaining the trust of farmers in the government is an essential third step.

Natural resource management and climate change

Asia's diminishing potential to meet expanding demand for water supplies and quality is a severe challenge that could reach crisis levels in many countries in the next couple of decades. Since much of the future water requirement is to be met from existing supplies, there is a need to change the institutional and legal contexts of water supply and use. Essential components of an agenda for action on water are improved water management through the removal of subsidies and taxes that lead to the misuse of water, the establishment of property rights, greater participation of water users in management, and the development of water markets whenever possible to send correct signals about the real value of this basic resource.

Climate change also exacerbates the stresses on agricultural production, particularly for lowand mid-latitude developing countries including those of Asia. In these areas, higher temperatures tend to adversely affect rice and wheat growth. Both mitigation and adaptation are important, requiring international capacity building and significant institutional innovation, such as specialized business services to low-income producers, locally accountable intermediary organizations, and simplified standards that are credible but still meet international standards.

Risk and insurance

Drought and climate change are just some of the risks facing the poor in rural areas. Other risks include pest outbreaks, changes in demand and world markets, illness, and death. Poverty is closely related to vulnerability. Better weather forecasting at the micro-level and timelier weather and market information dissemination to farmers could be a cost-effective way to reduce crop risks, especially in monsoon-dependent areas of highly seasonal rainfall. Moreover, Asia has been exploring ways of building innovative insurance markets around the risks that directly and indirectly affect Asia's rural poor, and it is important to develop cost-effective and sustainable insurance instruments. New technologies such as information and communications technology and innovations in terms of certifying and developing index-based insurance such as weather insurance may provide a real opportunity to mitigate the risks faced by farmers. More formal insurance mechanisms related to crop futures, including those organized or subsidized by the government, are currently underdeveloped in Asia and require careful assessment.

Institutions and community empowerment

Rural institutions mediate the access of the rural poor to assets, markets, technologies, and services, but institutions tend to be controlled by powerful elites. In order to empower communities, it is vital to organize poor farmers, workers, and small businesses, and to improve the transparency about and access to information on the functions, policies, and decisions of local government and rural development institutions and agencies. The former groups need to have clear oversight and monitoring functions in order to protect and preserve their own interests against the capture of power by vested interest groups. Organizations of the poor can help expand and strengthen access to credit, supply chains, technology, and other public goods.

Infrastructure

Investing in infrastructure, including road and water transportation and information and communication technologies, can help reduce spatial disparities and foster rural–urban linkages. However, the primary mode of transportation of many of the poorest is walking. Therefore, they will need some other intervention to allow them to benefit directly from investments in roads. In addition, transferring income in the short run from more-favored areas to less-favored regions must be supplemented by medium-term measures to facilitate migration through investment in education and subsidized resettlement.

Prioritization of Actions for Pro-Poor Growth

During the past few decades, the share of national budget allocations to agriculture has significantly decreased in Asian countries. Studies from India, China, and Indonesia suggest this reluctance to invest in agriculture is misguided because in general, agricultural spending has large effects on growth and some impact on poverty reduction. However, a "one size fits all" policy cannot be effective; instead, regional and country contexts should be taken into account. In Thailand, for example, education in rural areas may be essential to ensure that the rural poor who migrate and send remittances to rural areas have access to higher-skilled jobs; in Laos, however, investment in rural roads may be essential to ensure that the remote poor benefit from price liberalization.

Priority action areas for achieving inclusive agricultural and rural development include the following:

- Increasing technological Iinnovation capacity in new agricultural supply chains
- Investing in infrastructure and communications systems to reduce spatial disparities and foster rural–urban linkages
- Developing ecosystems services through public-private cooperation for meeting the challenges of water scarcity and climate change
- Facilitating the health benefits of agriculture and reducing the risks, especially the transmission of animal diseases
- Encouraging development of finance and insurance interventions for the poorest
- Providing effective safety nets and nutritional improvement in rural areas

Future policy research questions that need attention include:

- What policies and services are needed to provide asset-poor farmers with access to emerging market opportunities, such as high-value agriculture? This work should include attention to new forms of cooperation and contract farming as well as information services regarding prices, quality, and standards.

- What types of nonfarm activities especially those undertaken by the poorest)comprise the rural nonfarm sector, and what types of policies encourage the growth of these activities? This work should include new ways of strengthening rural–urban linkages as well as rural services and innovative rural financial institutions that embrace remittances and rural small- and medium-sized enterprises.

- What types of opportunities may evolve for biofuels in different settings in Asia, and what policies and regulations are needed for a balanced approach that may include smallholder farmers and small-and large-scale biofuel producers?

- What policies exist for climate change mitigation and adaptation in Asian agriculture? This work would include studies on carbon trading regimes and related programs and policies.

- What new policies are there for making water and irrigation systems in Asia more sustainable and pro-poor? This work would focus on the need for enhanced supply response in irrigated agriculture in different settings in Asia and emphasize water quality in the context of irrigation, which is so critical for the poor.

Conclusion

Agriculture and rural development strategies are a critical component of an inclusive growth strategy for Asia. Lessons can be learned from the region's recent history, but trends cannot simply be projected. The challenge is to determine what these lessons mean for the future. Developing a modern rural development strategy for poverty reduction in Asia in the coming decades requires recognition of institutional and technological changes, as well as of the growing role of a broadly defined and diversified agricultural system as opposed to the declining but relevant role of crop agriculture. Furthermore, the rural poor need to participate in the development and implementation of the relevant policies and programs, which need to ensure gender equity, inclusion of minorities, and participation of the poorest in markets and in the provision of services. All of this requires a fresh initiative for identifying and implementing the appropriate actions for inclusive agricultural and rural development in those regions of Asia that are home to the poorest and hungriest.

Chapter 9

Rural Diversity, Agricultureal Innovation Policies and Poverty Reduction

The purpose of this paper is to explore how agricultural innovation policies can better serve the goal of poverty reduction. The paper is based on a larger study commissioned by the World Bank. It focuses on the institutional factors that can favour poverty-reducing innovation in agriculture. The main premise is that agricultural innovation policies and programmes cannot start from agricultural research and extension and then hope to address poverty. This can only lead to standardised, 'one size fits all' solutions. The starting point has to be the analysis of the different types of poverty, their determinants and the contexts in which they occur, and the livelihood strategies that the poor implement to respond to their condition. From there we can move back to determine possible strategies for agricultural innovation, and only then can we look at the potential role of agricultural research and extension within each strategy.

The article emphasises that not all poor rural households are agricultural, nor is agriculture the best avenue for reducing income poverty for them all. Agricultural production can be a major element in poverty-reducing strategies, or it may play no role at all. The role of agricultural innovation in poverty reduction must be based on an analysis of the assets and context of rural poverty in specific locations and on an understanding of both the direct and the indirect effects of increases in agricultural productivity on different classes of poor people. Section 2 of the paper reviews the nature of rural poverty, emphasising its complexity and the relative importance of farm and non-farm activities. Section 3 examines the effects of agricultural innovation on poverty; its direct effects include increased production and income for farmers, while indirect effects include lower food prices and increased employment opportunities. Section 4 reviews the concept of agricultural knowledge and information systems and examines the changing institutional scope of agricultural innovation. Section 5 presents three sets of strategies for rural poverty reduction, based on an understanding of the asset and contextual base of different classes of rural residents. These strategies differ significantly in their emphasis on different types of agricultural development for poverty reduction. Section 6 summarises the argument and presents conclusions.

Rural Poverty of Rural Poverties

According to World Bank estimates, 1.2 billion people lived in absolute poverty in 1998, depending on an income of less than US$1 per day. An additional 1.6 billion lived on less than $2 per day. The number of people in the former category has remained constant in the last decade, while there are now an additional 250 millions living on less than $2 per day. Around two-thirds of the world's poor live in the rural areas of the developing world, and in sub-Saharan Africa they can constitute as much as 50 to 90% of the population. It is estimated that in the late 1980s there was a total of nearly one billion poor rural people in 114 developing countries. While there is a broad consensus that agricultural development cannot by itself overcome the state of deprivation of so many people, there is also little doubt that without the long-term and significant growth of the agricultural sector, there would be fewer opportunities for significantly reducing rural poverty.

One reason is that in the year 2000 almost 60% of the total population of the developing countries lived in rural areas.

Poverty and its determinants

Poverty is multi-dimensional. Its causes are diverse, its manifestations and definitions are contextual, and it is not only a state of deprivation but also a set of processes. Rural people, their goals and the livelihood strategies they adopt to achieve them, are very diverse. Their livelihood strategies are aimed at increasing income, reducing vulnerability, improving well-being and ensuring food security. These outcomes are explained by the asset position of households and the characteristics of the context where assets are used including institutions, power structures, markets and policies and their organisations, trends and shocks.

Access to land is a major determinant of the livelihood strategies of rural households. The highly uneven distribution of land is a major reason for rural poverty in many countries of Asia and Latin America, while, according to a paper published by the International Monetary Fund, in sub-Saharan Africa, 'the poor quality of land and the erosion of customary land rights have become the major obstacles to agricultural growth and alleviation of poverty'. In India, about 70% of rural households own less than one ha of land, 11% are landless, and 24% do not operate any land at all, sometimes even when they own it. In sub-Saharan Africa and Asia, one-third of smallholders subsist on plots too small to support their families. In the more agriculturally favoured parts of Nepal, 40% of landless or almost landless households are poor, while only 18% of medium and large farmers are in this condition; in the mountains, poverty incidence among the same groups is 77% and 24% respectively. In Mexico, access to land is the most important determinant of total rural household income.

Access to irrigated land has a particularly strong effect on total income. Human capital assets are another major determinant of the livelihood strategies of rural people. In Mexico, the number of years of education of the adult members of the households has a strong positive effect on total income. However, this study also concludes that access to education has a higher pay off in the non-agricultural rural labour markets, and in fact has a negative effect on agricultural income because educated household members seek employment in other sectors of the economy. In Ghana education plays a role in increasing household welfare only after completion of primary or secondary school. In China, increased literacy due to higher investment in education has made the largest contribution to rural poverty reduction.

Rural infrastructure directly and indirectly affects the livelihood strategies of rural households. In China, improved rural roads, electrification, telecommunications and irrigation infrastructure have significantly contributed to the reduction in poverty. In India it has been shown that public investment in roads has made a major contribution to poverty alleviation in some of the less favoured agroecoregions. Given that their assets and contexts are so heterogeneous, it is no surprise that rural households pursue a number of different livelihood strategies.

Agricultural production can be a major element here, or it may play no role at all. The notion that all rural households are agricultural or that agriculture is the best way of reducing income poverty for them all should be discarded as untrue and unrealistic. The need to diversify rural employment and income is increasingly important in all developing regions in the world. Based on a review of a large number of national and sub-national studies, Reardon et al. estimate that non-farm activities account for 36% of total rural income in West Africa, 45% in East and Southern Africa, 35% in East Asia, 29% in South Asia, and 40% in Latin America.

In rural China, employment in the non-farm sector grew from 7% in 1978 to 29% in 1997. In 1997, 36% of rural income came from non-farm sources, although agriculture accounted for up to 90% of total income in the less developed areas of the country. In 1997 rural non-farm enterprises accounted for more than 25% of the national GDP, up from nearly zero in 1978. Without this growth of the rural non-farm economy, China's GDP growth rate would have been lower by 2.4% per year. When the only source of income is agriculture, household income correlates strongly with land ownership. However when non-agricultural livelihood sources are available, this link is broken and it becomes possible for income inequality to be reduced even if land holdings are unequal. In Egypt, Adams has shown that non-farm income is of great importance to the poor. It contributes almost 60% of their total income and is the most significant contributor to the reduction of inequality.

The agricultural path out of income poverty is directly relevant only to some of the rural poor: those who have access to sufficient land and other assets needed for agricultural production, and who operate in contexts that provide the correct incentives for those assets to be sufficiently productive over long enough periods of time, to lever their households out of poverty. Attempting to force households and communities who lack these resources and contexts to base their development on agricultural production, is to push them deeper into a poverty trap.

Direct and Indirect Effects of Agricultural Innovation on Poverty

Agricultural innovation can have both direct and indirect effects on reducing poverty. Which is more important will be determined largely by the relative speed with which a household adopts new technologies, by the status of the household as a net food buyer or seller, by the degree of market liberalisation conditioning whether particular products are tradable or non-tradable, and by the institutions and incentives facing farmers.

Direct effects

The direct effects of technological innovation on poverty reduction are those benefits captured by the farmers who actually implement the changes, and they manifest themselves in the form of higher profits. New technologies can improve a farmer's income when they reduce the marginal cost of producing one unit of output. Since output prices will for a time be driven by the prevalent technology, profits will increase for those who adopt the new one, with early adopters benefiting the most. Eventually all or many farmers may adopt the new technology, causing increases in output and a possible reduction in output prices. The profit margin created by the new technology may or may not disappear completely. Late adopters or non-adopters who continue to produce with the old technologies when costs and prices are already determined by the new ones may be negatively affected.

If this occurs in a closed economy, or in a region which is in effect protected due to poor access or any other variable, the adverse effect from the producer's point of view of a new productivity-enhancing technology on farm prices will be faster, and there will be a higher premium on early adoption. If the process takes place in an open economy, local innovation will have a very small or even negligible effect on reducing the price of agricultural products, but increased productivity will reduce costs per unit of output, and adopting farmers will benefit from larger profit margins.

However, farming in an open economy means competing on a global scale, with output prices determined by the most productive. Poor farmers are not usually among the early adopters; they lack the necessary access to information, capital, skilled labour, roads, and other such factors. It follows that they stand to gain much less than larger, commercial farmers from the direct effects of technical change, particularly in open economies. Millions of such poor farmers in recently liberalised economies are having great difficulty matching the unit costs of production of their counterparts working under better conditions. In a very real sense, they are running behind international prices determined by the higher productivity of farmers in parts of the world with better comparative advantages, due to more favourable asset positions and production environments, better technologies and policies and institutional incentives more conducive to success.

Indirect effects

The indirect effects of technological innovation on poverty reduction are the benefits passed on to others by the farmers who actually implement the changes. These effects can take one or more of several forms: (a) lower food prices due to higher agricultural productivity and output; (b) employment generation in agriculture; and (c) broad-based economic growth through production and consumption linkages with the non-farm economy.

Lower food prices

Lower food prices are an inevitable consequence of increased productivity due to technical change. Recent studies have established that in the absence of Green Revolution technologies, food crop prices would have been 27 to 41% higher over the past 25 years. Lower food prices are a fundamental contribution to increasing the welfare of the 300 million urban people who live in absolute poverty and who spend very large proportions of their meagre income on food. In India, the poorest two-thirds of the population which include the majority of rural people spend 73% of their income on food.

Lower food prices also have a major impact on the vast majority of the rural poor, who are net buyers of food either because they are landless, or because they have insufficient land to meet household consumption needs. Eleven percent of India's rural inhabitants have no access to land, and 27% do not operate a farm. In Mexico, only 28% of the peasants in the 'land reformed' sector are net sellers of maize, the main food staple of the rural population and main component of the traditional peasant farming systems. In Nicaragua, 23% and 28% of rural households are net buyers of maize and beans respectively, while only 39% and 37% are net sellers of these two major food crops. Because of this, it has long been argued that the main impact of agricultural innovation on both urban and rural poverty should come from increased productivity resulting in lower food prices. As in the case of the direct effects, market liberalisation alters the importance of indirect effects on poverty through food prices. The price of food paid by the urban poor and by rural net food buyers will largely be defined by global trends rather than by what occurs at a local or even national level. Whether a country with a liberalised economy is self-sufficient in a particular tradable agricultural product will mean little in terms of its average price to consumers.

Agricultural employment and wages

Some improved agricultural technologies can increase total on-farm employment, particularly when they stimulate agricultural output per unit of land per year. Depending on the conditions of the labour market, this can result in increased wage rates. Otsuka et al. and Otsuka studied the effect of agricultural innovation on labour demands in the Philippines and other Asian countries. While the first-round effect of adopting modern varieties of rice did increase labour use both because of greater yields per cropping season and because of multiple cropping, they also found that such innovation soon lead to the adoption of labour-saving technologies agricultural machinery in particular, which more than offsets the gain. In the Philippines, for example, the average use of labour per hectare of rice declined by 20% between 1985 and 1998.

After reviewing the evidence on the impact of technological change on the labour markets, Renkow concludes that 'although rising real wages might appear to be the obvious result of increased labor demand caused by technological change, empirical confirmation for this is small. Rather, available evidence indicates stagnation or, at best, small increases in real wages.'

Linkages with the non-farm economy

There can be production linkages between the farm and non-farm sectors. Upstream linkages are those stimulated by growth in the farm sector, inducing the non-farm sector to increase its activities to supply inputs and services to the farm sector. Downstream linkages arise when the non-farm sector is induced to invest in capacity to supply agro-processing and distribution services, using farm products as inputs. An analysis of these linkages is beyond the scope of this paper, but the reader is referred to Reardon et al., 2001 and 1998.

Household market integration and the impact of direct and indirect effects

The net impact of direct and indirect effects on poverty reduction will depend to a large extent on the influence of three factors: (a) the tradable status of the product; (b) the position of the household in the market as net buyer or net seller of the product; and (c) if the household is a farming family, its position as an early, late or non-adopter of the technology.

The effect of agricultural innovation on food prices is highly dependent on the tradability of the commodity in question. As a general rule, when a commodity is nontradable, increases in productivity will have a faster and stronger effect on the price of outputs, and consumers will tend to benefit most, followed by those farmers who are among the early adopters. Late or non-adopters which often include most poor farmers will benefit least or even lose, although they may benefit if they are net consumers of the particular product.

On the other hand, when a commodity is tradable, local prices will reflect international prices. Local increases in productivity will not have a significant effect on lowering food prices, and thus adopting producers stand to gain while the effect on consumers is neutral.

While a commodity may be tradable in the world market or in the urban centres of a given country, it can also be non-tradable elsewhere because of poor roads, long distances, or cultural food preferences. In regions such as Latin America, West Asia and North Africa where many or most of the poor are located in urban areas and where local prices reflect international prices, the indirect effect of local gains in productivity due to technical change in agriculture will be low, but the direct effect on the farm income of adopting net producers will be high. In countries such as many in sub-Saharan Africa and some areas of South East Asia where most of the poor live in areas under *de facto* protection due to poor roads and high transaction costs, the indirect effect on food prices of local gains in productivity due to technical change in agriculture will be high for consumers, but the direct effects on the farm income of net producers will be low.

From the point of view of poverty reduction, the relevant price is determined by the idiosyncratic position of each household *vis-à-vis* that market. For net buyers of food, the relevant price is the market price plus the transaction costs of purchasing. For net sellers it is the market price net of the transaction costs of selling. Households that neither sell nor buy a particular crop find themselves with a farm gate implicit price that is too low to sell and too high to buy.

The process of technical change can create losers, or winners and losers simultaneously. As Kerr and Kolavalli point out, the distributional effects of technical change (among different types of farmers, between farmers and wage labourers, between producers and consumers, among regions) will depend on policies and institutions. De Janvry and Sadoulet add that trade-offs between the direct and indirect effects of agricultural technology on poverty reduction will be higher when there are institutional gaps or barriers that discriminate against the poor in their access to public goods, when there are large market failures, and when land and other assets are unequally distributed.

Thus it does not make sense to select one single strategy *a priori* as the most convenient to reduce poverty. The constituency of poverty-reducing agricultural innovation includes the urban poor, small farmers who are net sellers of their outputs, small farmers who are net buyers of food, agricultural wage workers, the rural nonfarm wage and self-employed, and, of course, the very large number of those whose livelihood strategies combine elements of several of the above. The correct mix of policies and instruments has to be determined on a case-by-case basis. Standardised and universal recipes, such as 'prioritising poor farmers in marginal regions' or 'prioritising resource-rich farmers in high-potential areas' is not what is needed if the goal is to improve the impact of agricultural innovation on poverty. More likely than not, what developing countries need are comprehensive sets of differential policies, each aimed at addressing specific populations and objectives. A problem of this approach is that designing, managing and implementing such policies is much more difficult and demanding than a 'one size fits all' strategy.

Institutional Changes in the Agricultural Knowledge and Information System of Developing Countries

The concept of agricultural knowledge and information systems was coined by Röling . According to FAO and the World Bank 'an Agricultural Knowledge and Information System links people and institutions to promote mutual learning and generate, share and utilise agriculture-related technology, knowledge and information. The system integrates farmers, agricultural educators, researchers and extensionists to harness knowledge and information from various sources for better farming and improved livelihoods.'

An essential element in the AKIS concept is that it views agricultural research and extension as necessary but, by themselves, insufficient elements in complex innovation-oriented institutional arrangements. The concept clarifies the distinction between agricultural research and extension, and innovation and technological change. As Anderson has pointed out, it is not correct to attribute all the effects of technological change to agricultural research and extension. The focus is not on research or on extension *per se*, but on *innovation* and on the institutional arrangements that can favour it. This is a sharp departure from the conventional view of innovation as a linear and rather mechanistic process started in highly skilled and specialised organisations conducting basic and strategic research, then moving down the line to applied research, adaptive research, technology transfer, extension and, finally, farmers as passive adopters of knowledge and information generated elsewhere. As Röling and Jiggins have stated: 'It has become common practice to speak about 'agricultural knowledge systems', i.e., to use a systems approach for looking at the interaction among the actors operating in a 'theatre of agricultural innovation'. Innovation emerges from this interaction and is no longer seen, as was customary in the 'transfer of technology perspective', as the end-ofpipe product of a sequential process.'

Agricultural research, extension and or development projects are important ways of stimulating agricultural innovation. However, only a fraction of the world's poor are directly involved in such projects. Hence, while it is very important that they are well designed and managed so that they can better alleviate poverty, their largest impact will be *indirect* and will include the diffusion of innovations propelled by factors not under the direct control of any project. If we are concerned with reducing poverty, particularly rural poverty, on a scale compatible with its global magnitude, we must emphasise policies and processes with wider effects at national and international levels.

Should the public sector invest in agricultural research and extension

Agricultural knowledge and information flow through a variety of channels, including private sector firms, universities, NGOs, agro-industries and commercial suppliers of agricultural inputs and equipment. According to the concept of subsidiarity of the state, public action is justified when market failure prevents the provision of a socially optimal level of a good or service, and or when there are externalities associated with that good or service. In the case of rural poverty reduction, the key argument is that, in the absence of public intervention, a large number of potential users will be denied access to agricultural knowledge and information which is readily available to other, non-poor farmers. Most people would agree that levelling the field in this respect is a basic public responsibility. Also much of the knowledge and information generated and disseminated by research and extension is in the nature of public goods.

A more complex environment

The environment in which agricultural research and extension operate today is increasingly more complex than in the days of the Green Revolution.

Decreasing public funding for agricultural research and extension

During the 1960s and 1970s, investment in research grew rapidly, at an estimated global average of over 6% per year. During the 1980s, investment stagnated or declined, particularly in Africa and Latin America, influenced not so much by new public policies *vis-à-vis* research, but more by overall efforts to cut public spending as part of the structural adjustment programmes of the 1980s.

The decline in public funding caught the national research organisations badly prepared, as in many countries the research institutes' administrative costs and number of scientists had grown faster than the budget for new investments and operational expenses. In Latin America, for example, between 1981–5 and 1992, the number of researchers grew by 22%, while the total budgets decreased by 15%. Many of the better qualified scientists left the public institutes as their real salaries dropped and the possibility of doing effective research declined.

Similar trends were experienced by the international institutes of the CGIAR, which suffered important budget cuts at the same time as the number of centres – and the demands placed on them increased. In the 1980s, these trends were even more drastic in the case of extension. In Latin America, many national governments simply abolished their extension services, which were widely seen as having become grossly inefficient, bureaucratic and ineffective.

The decline in public funding has been only partly offset

by the growing importance of private research efforts, which by the late 1990s accounted for only 10 to 15% of the total investment in developing countries. Moreover, much of this private research is concerned with commodities and technologies of little importance to small and poor farmers.

The result is that in many developing countries, the current calls for greater research and extension support for poverty reduction policies are constrained by the lack of research and extension systems capable of rising to the challenge. In many countries, efforts to revive these national organisations through internationallyfunded programmes, have led nowhere.

Greater institutional complexity

At the same time as the official research and extension agencies were experiencing the decline described above, new institutional actors began to make an impact in developing regions. These include private firms, NGOs, universities and research institutes, foundations, farmers' organisations, new ministries for environment, social welfare and science and technology, agroindustries, and, more recently, local governments at the regional and municipal levels. Some of these have become involved directly in the organisation and delivery of agricultural knowledge and information, while others have played a major role in the formulation and advocacy of policy. A recent study of four rural districts in Kenya, for example, reports the active involvement of over 30 different organisations in each district, including local community groups, seed suppliers, NGOs, traders, plus official research and extension programmes and institutes. All of them provide farmers with services of direct importance for agricultural innovation at the local level.

Unfortunately many studies conclude that these numerous agencies often act with little or no coordination or even contact among themselves. Each promotes its own agenda or pursues its private objectives. Farmers' and community organisations often lack the skills, power and resources to take control and manage this complexity on behalf of their members. It is not only that there are now many more actors involved in agricultural innovation in developing countries, but also that the mechanisms for planning, designing and delivering services are themselves also more diverse. In many countries the financing and delivery of extension services are now separate functions, for which different institutions are responsible. In some areas private agro-industrial firms contract production out to small and medium farmers; sometimes these private firms sign specific contracts with government agencies or universities to conduct applied and adaptive research to solve specific bottlenecks. NGOs often formulate their own programmes and projects, taking advantage of a very well developed networking capacity. On other occasions, NGOs act as subcontractors for official programmes, or develop joint venture agreements with them. Governments in many Latin American countries have established foundations that act as quasi-private organisations. Liberalisation has opened up a market for small and medium private entrepreneurs and local community organisations to take charge of input supply and technical advisory services.

Funding mechanisms for agricultural research and extension have also evolved. The old ways of block grants or regular financing through the national budgets are giving way to an array of procedures that share the common characteristic of linking funding to the delivery of specific and well defined outputs. Competitive funds are increasingly being used in Latin America, Asia and Africa for this purpose. While the move to emphasise delivery of outputs is healthy in many situations, it has also resulted in a significant dispersion of research efforts, lack of strategic goals, exacerbation of competition for funds, weakened cooperation among different agencies, and dwindling support for long-term investment in research facilities and human resources. Of particular concern is that in practice these competitive funds often discriminate against the poor and the marginal regions, simply because the supply side of the equation tends to be weaker compared with that for dynamic regions, profitable and competitive commodities, and commercial farmers. National research agencies are also attempting to commercialise research results and services, applying intellectual property rights when needed. Most Latin American agricultural research institutes now have procedures in place or have developed subsidiaries to patent and charge royalties for their products or to establish contracts with private firms to evaluate inputs, conduct soil tests, or multiply seed.

As governments push research agencies to generate an increasing share of their budgets from these commercial operations, problems important to small and poor farmers and marginal regions are often ignored since they offer fewer opportunities for commercialisation.

In summary, while all of these developments have helped to resolve at least some of the many important shortcomings of the traditional agricultural research and extension systems, and while the great variety of agencies involved contributes to the richness of a country's agricultural knowledge and innovation system, one cannot lose sight of the fact that the poor are very often worse off. The emphasis on shorterterm results, on better rates of 'success', on greater financial self-sufficiency, on relatively simple problems that fit within the framework of three- or four-year projects, on a greater capacity from the supply side to formulate and prioritise problems and place them in project formats, on improved ability to seek information about multiple options and negotiate with multiple potential partners constitute the new 'rules of the game' for agricultural research and extension these days, and they make it very difficult for the poor to 'play'.

Targeting the poor in research and extension

Partly in response to the above trends and their effects on the poor there has been a move to develop new approaches to prioritise and target the needs of small farmers in agricultural research and extension. Byerlee discusses different approaches to improve the targeting of agricultural research for poverty reduction. The basic economic surplus models used by many national agricultural research organisations to prioritise research can be refined to differentiate between different categories of consumers typically, income or expenditure groups and producers normally represented by classes of farmers according to farm size. Also, geographical information systems are used to add a regional perspective to the allocation of research, and this can relate to poverty alleviation if there are differential spatial distributions between the poor and the non-poor. According to Byerlee 'the effectiveness of targeting will depend on the extent that: (i) benefits of research are captured by producers; (ii) poor producers depend on agricultural incomes; (iii) poverty is correlated to the targeting variable, (iv) the targeted research can generate economic surplus higher than the cost of the research; and (v) the degree and nature of technological spillovers.'

Demand-led, participatory targeting and prioritisation approaches are being increasingly applied in agricultural research, extension and development. In this area there has been an explosion in the number and variety of methods and tools, from the now more traditional participatory technology development and rapid rural appraisals, to the promotion of local farmers' research committees and farmers' field schools, participation of farmers and farmers' organisations in the boards of the research and extension agencies and the relatively recent development of competitive agricultural technology funds.

These participatory, demand-led approaches have provided practical, effective and cost-efficient solutions to the very complex problem of how to make agricultural research, extension and development more relevant to the needs of poor farmers. However, they are not free of limitations. Pretty for example, estimates that in the mid-1990s there were about two million farmers, in 20 countries, involved in projects promoting participatory sustainable agricultural technologies and processes, and yet he states that these are 'small islands of success' and that 'there remains a huge challenge to find ways to spread or scale up' these processes. Farmer participation does not always ensure that the poor lead in influencing the agenda of these programmes; local rural elites tend to control decision-making, as in the case of the local farmers' research committees promoted by the International Centre for Tropical Agriculture and others in Honduras, where illiterate farmers are underrepresented in local organisations. Finally, it must be said that these types of participatory methods are most appropriate when the purpose is to maximise the direct effects of agricultural innovation on participating individuals, households or communities, but are less relevant in contexts in which indirect effects are important.

Differential Strategies to Impact on Rural Poverty

In the previous pages we have argued that the net impact of agricultural innovation on poverty will be the result of: (a) the livelihood strategies of the households, which in turn are conditioned by the interaction between their assets (capacities) and their contexts (incentives); and (b) the relative impact of the effects of agricultural innovation on the net income of farming families, food prices and agricultural and non-agricultural employment.

The interplay of the above elements suggests that differential strategies are required to build situationspecific AKIS relevant to particular combinations of these factors. (A) Market-driven AKIS that impact on poverty primarily through lower costs of food for the urban poor and for rural poor who are net buyers of food, through more and better employment opportunities, and through production, consumption and investment linksto stimulate the non-farm rural economy (B) Market-oriented, asset-constrained AKIS, that may impact on poverty through direct and indirect effects (C) Context- and asset-constrained AKIS, where agricultural innovation can make limited direct contributions to poverty reduction and where nonagricultural livelihood strategies predominate.

The first strategy is relevant to situations in which agricultural innovation is spurred by favourable contexts and asset positions. In sector agriculture is a profitable and competitive enterprise. AKIS develop primarily due to the action of market forces, and specifically because of the profit opportunities found by commercial farmers and private firms and entrepreneurs in the services and agroindustrial sectors. The profit motive is behind the linkages and interactions in this strategy. Research and extension find here near optimal conditions to display their capabilities. Private research naturally targets these type-A situations. Commercial farmers have the skills, education, networks, organisations, political power and capital required to mobilise and influence both public and private agricultural research and extension when and where necessary.

Direct effects on poverty are of little importance in this situation, as few of the farmers operating in these conditions are poor to begin with. On the other hand, it is under these conditions that indirect effects are maximised: high adoption rates result in rapid improvements in productivity, driving food prices down on a global scale. When intensive agricultural systems are the norm, thousands of jobs can be created, and these areas are characterised by large seasonal migration of farm workers from less favoured regions, often across countries. By definition, these agricultural systems are linked in many ways with the non-farm economy, as their requirements for transportation, supplies, professional services, marketing systems and processing are very high. Typically high-income farmers and dynamic farms are also investment-linked with nonfarm enterprises.

Is there a role in developing countries for public policies and interventions directly related to agricultural innovation under such conditions? The answer is a resounding yes. But these policies and interventions will tend to be of a horizontal nature, the kind that generally create or strengthen institutional frameworks. These public policies and interventions should have the goal of exploiting the intersection between public goods and private interests, by:

- Developing clear regulatory frameworks and strengthening intellectual property rights
- Protecting these agricultural systems from unfair trade competition and non-trade tariffs and barriers to commerce, as well as promoting private-public cooperation for the purpose of opening new international markets
- Developing national science and technology policies that do not discriminate in favour of any one particular kind of research organisation but that, on the contrary, promote diversity, specialisation and competition, all of which will lead to cooperation when necessary, in particular if the private sector is provided additional incentives through such means as competitive science and technology funds or specific research contracts
- Supporting through direct public investment basic and strategic research in those fields of science and technology that are most relevant to agricultural systems with actual or potential competitive advantage
- Funding the specialised training of highly skilled young scientists
- Promoting the establishment of cooperation agreements between national universities and research institutes and their counterparts in other countries
- Stimulating linkages and cooperation for innovation between different actors in agro-industrial and marketing chains
- Investing in public but non governmental organisations such as foundations endowed with public and private funds that manage venture capital for new enterprises to bring new technologies into the country
- Adopting policies that create favourable conditions for foreign investments in R&E and in innovative for-profit enterprises.

If the correct institutions and policies are in place to ensure that the benefits of growth and innovation are broadly shared in a society, all the above public policies and interventions will lead to win-win scenarios for growth and poverty reduction. In addition, if successful in stimulating greater private investment in research and extension, these policies and instruments could free public resources for investment in other situations.

Market-oriented, asset-constrained AKIS

Zone B in Figure 1 depicts a situation in which small farmers, who may have the incentives to embark on market-oriented agricultural innovation processes, are predominant but lack the capacity to respond fully to this favourable context, either because their assets are too limited, the productivity of such assets is low, or because the transaction costs they face are too high. For reasons that will be discussed in the following paragraphs, it is likely that this group of small family farms represents the best opportunity in economic, social and also political terms for linking agricultural innovation and poverty reduction policies in developing countries.

Effective type-B policies and programmes will yield both direct and indirect impacts on poverty. Many of the farmers in type-B situations are poor, and this opens space for direct effects on the net income of farming households. Second, all the research on the non-farm rural economy shows that it is in this type of situation that farm non-farm linkages develop best and have the largest effects on the welfare of rural communities.

Third, small farmers in type-B situations produce a large share of the world's food, as in the notable case of rice in Asia or vegetables for the domestic market in much of Latin America, or milk in Africa. The market will not produce these results by itself, otherwise it would have already done so. Neither will results be achieved in the absence of clear and viable links to markets. Strategies for type-B conditions must have a clear and strong market orientation, for which proactive public policies are needed. In type-B situations, one often finds the condition described by Röling and Jiggins namely that there is potential for AKIS to develop, but that a lack of synergy among the different private and public agents constrains its realisation.

Public policies and interventions under type-B conditions are needed to: (a) increase the assets available to small farmers, as well as their productivity; (b) lower transaction costs and diminish the institutional constraints and the market failures that hamper the productivity and innovation potentials of family farms; and (c) promote the development of an effective AKIS by stimulating synergistic interaction between public and private agents, including farmers. What types of public policies and interventions can be successful in this context.

- Supporting efficient and effective NARS, understanding that the concept must not be restricted to the conventional agricultural research agencies, but should apply also to universities, private R&D firms, NGOs, farmers' organisations, foundations, etc. While action on the demand side e.g., competitive funds targeted at small-scale farmers is of great importance, there is also a need for policies and interventions that more directly target the development of capabilities on the supply side. These include investment in the training of the NARS's staff and improving their management systems. Demand-driven approaches do not work well in the face of weak suppliers of agricultural research and extension services.

- Revitalising extension and farm advisory services. This means changing their outlook from primary production to a market orientation, empowering users so that they exert ownership over the extension services, decoupling public financing from private service delivery, decentralising decisionmaking and involving local governments.

- To the extent that it is feasible at any given time and

place, give priority attention to non-food crops e.g., cash crops, boutique and niche products, nontraditional exports, crops that can be produced under contract with agro-industries, as well as rural farm-based enterprises that generate new types of goods and services demanded by the growing urban sectors, such as agro-tourism and the management of rural areas for recreational purposes. Of course, large-scale emphasis of non-food staples will be practical and feasible only when the country has the foreign exchange required to purchase basic food staples on the world market, and when the infrastructure and the marketing and processing sectors are developed to the extent needed for the services-intensive production of these kinds of agricultural products.

- Framing agricultural innovation within broader policies and programmes aimed at the broad-based development of local rural economies. In particular, agricultural research and extension should strive to link small-scale farming with the non-farm economy, by prioritising those enterprises, processes and production and post-harvest technologies that: (a) are more intensely linked with rural services and industries; and (b) save labour from on-farm work so that it can be invested in non-farm self- or wageemployment, wherever these opportunities exist.
- Promote small farmers' economic organisations as institutional platforms for collective action to confront problems of barriers to market access, economies of scale, and transaction costs.

Context and asset-constrained AKIS

Sector C in Figure 1 represents a situation in which households lack most types of assets aside from unskilled labour and, sometimes, a very little land e.g., less than a hectare, and, at the same time, operate in unfavourable environments. In short, the potential for agricultural development resulting in sustainable reduction of poverty levels is very limited or non-existent. While agricultural development projects can create artificial systems of incentives for a certain period and for a limited fraction of these millions of households, the fact remains that often these regions lack growth engines i.e., regular and dynamic sources of demand either for labour, services or products to provide the stimuli needed for long-term and widespread poverty reduction. As discussed in previous sections, those type-C households with access to land often engage in subsistence farming because: (a) they lack better employment options; (b) they have developed diversified livelihood strategies in which agricultural production complements other sources of income, often from unskilled agricultural labour, remittances and subsidies or from what Reardon et al. call refuge rural non-farm activities; and (c) transaction costs are so high as to effectively bar them from operating in the market as sellers and or as buyers of most agricultural products.

In type-C situations, poverty reduction policies and programmes must be even more broad-based than in type-B scenarios. Given the very limited endowment of agricultural assets, even significant long-term increments in agricultural productivity will usually have a very small impact on total household income. Farming systems under these conditions are very often in a state of finely tuned equilibrium within their very constraining contexts, which means that unless the constraints are removed or at least significantly reduced by means of broad-based development policies, there is little room for improvement.

Broad-based development policies aimed at type-C situations will normally include interventions aimed at trying to move these households towards a type-B condition, by enhancing their asset position e.g., land distribution, access to credit, training, education, and health programmes, strengthening community organisations, and or by improving the context in which they operate e.g., roads, irrigation, improving the capacity of local governments, supporting more efficient markets. In the absence of such changes, the potential for agriculture-based development will remain very low.

Does this mean that agricultural research and development cannot play a role under type-C conditions? Of course not. But successful efforts will tend to be based on an understanding that, under type-C conditions, selfemployment in agriculture will often be only one element many times not even the most important one of diversified household livelihood strategies.

Many people call for a greater share of agricultural R&D resources to be targeted on regions characterised by marginal environments for agricultural production. Their arguments include considerations of public versus private roles and of the efficiency of public investments if the private sector largely takes care of type-A areas and farmers, this should free public resources for investment in agricultural R&D in areas, crops, problems and populations not being served by private agricultural technology companies. The equity argument is based on the notion that the vast majority of the rural poor are located in marginal areas and, conversely, that there are few rural poor in the more favoured agricultural environments. While this is true in the sense that most poor rural people do live in marginal areas, there is also a very significant number living in more favourable environments.

While these arguments show that there is certainly some room for agricultural innovation to contribute through direct effects to a reduction in income poverty, they do not dispel the findings of many other studies that conclude that the poorest of the poor, in the worst environments, almost always develop diversified livelihood strategies to try to improve their well-being and their income.

What types of agricultural innovation policies and interventions can maximise the contribution whatever its magnitude of agriculture to poverty reduction in type-C situations? Recent research has shown that, while local networks of social capital play important insurance and solidarity functions, they can promote significant movement towards reducing poverty only when they cut across traditional family, clan, tribal or local community groups and link to outside networks. In other words, successful innovation in type-C situations depends on building local institutions, networks and organisations that help mobilise the very scarce resources of these communities, and link them to external networks. The strategies that have shown the greatest potential for spurring institutional innovation for agricultural development under type-C conditions include participatory technology development, on-farm adaptive research by NGOs and local organisations, and farmer-to-farmer extension systems. In Peru and Bolivia, the International Fund for Agricultural Development has experimented with programmes that transfer cash to local organisations empowered to hire technical assistance services as needed, with minimal external intervention. These organisations often enter into risk- and profit-sharing agreements with their advisers to develop new marketoriented enterprises. Also, in the Andean region of Latin America, many municipalities with high proportions of indigenous populations and with active NGOs, have seen the rise of *'mesas de concertación'* which are institutional arrangements for developing private-public medium term action plans for local development, in which agriculture and natural resource management projects often play a very important role. The biggest challenge yet unresolved is to find a way in which these local experiences can be extended so that their impacts are commensurate with the magnitude of one billion poor rural people.

In type-C situations, formal research and extension organisations will probably be effective if they operate through agreements with local organisations and NGOs able to facilitate local-level innovation processes, with the research organisation playing a technical and methodological backstopping role, as CIAT has done in Colombia and Central America.

Conclusion

Poverty is multi-dimensional. Its causes are diverse, its manifestations and meanings are contextual, and it is not only a state of deprivation but also a set of processes. Failure to recognise these facts has led to a succession of fads of simplistic 'one size fits all' approaches. If we want to improve the performance of agricultural knowledge and information systems *vis-à-vis* poverty, it is time that we learned to deal with its diversity by means of customised approaches.

2. When the goal is poverty reduction, agricultural innovation policies and programmes cannot start from agricultural research and extension and move towards poverty. The starting point has to be the analysis of the different types of poverty rural and urban, their determinants, the contexts in which they occur, and the livelihood strategies that the poor implement to respond to their condition. The more we move into conditions of poverty, deprivation, and unfavourable environments, the greater the need for these types of broad-based approaches. Only in certain circumstances will research and extension find themselves in the driver's seat. Even then, to impact on poverty, R&E will have to be framed in the context of broadbased development policies. They will fail if they try to tackle this issue in isolation.

Given how heterogeneous the assets and contexts of the poor and the determinants of poverty are, it is no surprise that rural households pursue a number of different livelihood strategies. Agricultural production can be a major element in such strategies, or it may play no role at all. The notion that all rural households are agricultural, or that agriculture is the best way of reducing income poverty for them all, must be discarded as untrue and unrealistic. The agricultural path out of income poverty is relevant only to those who have access to sufficient land and other assets needed for agricultural production, and who operate in contexts that provide incentives for such assets to be sufficiently productive over long enough periods of time to lever the household out of poverty.

Under the increasingly prevalent condition of liberalised agricultural markets, those farmers who are early adopters of new technologies and who can keep pace with continual innovation, stand to gain from the direct effects of technological change. In the long run, many poor farmers in developing countries will only be able to benefit from the direct effects of agricultural innovation if they operate under conditions of *de facto* or *de jure* protection from international trade, or if public policies are put in place to allow them to significantly increase their productivity and or to diversify into production systems in which they have a competitive advantage.

Indirect effects of agricultural innovation through lower food prices are a fundamental contribution to increasing the welfare of the urban poor and also to a vast proportion of the rural poor who are net buyers of food. There is little hope for progress in the fight against poverty on a global scale if agricultural research and extension diminish their overall support to those regions and farmers that can create this type of indirect effect. However, the rise of privately-funded R&D in developing countries offers a new set of conditions for defining the contribution of the public sector to agricultural innovation which targets these types of indirect effects on poverty. There are now much greater opportunities for the public sector to reduce its direct involvement in implementing the research and delivery of extension while at the same time enhancing its capacity to stimulate and support private efforts, and to take care of those objectives that are outside the scope of private interests.

An additional indirect effect of agricultural innovation that needs to receive much greater attention than in the past is the growth of the rural non-farm economies. All the recent evidence shows conclusively that growing numbers of rural households derive an increasing share of their total income from this source. Non-farm income is of particular importance to the rural poor, including many agricultural households. Some traditional concepts need to be sharply revised, such as that which states that pro-poor agricultural technologies must be labour-intensive, as if the poor do not have other, better remunerated employment opportunities. Agricultural research should actively seek and exploit opportunities for developing and strengthening production, expenditure and investment linkages with the non-farm economy.

During the last two decades of the twentieth century, the AKIS of developing countries have undergone significant institutional changes. The budgets of the traditional government research and extension agencies have tended to decrease, often accompanied by a more general weakening of the organisations' capacities. At the same time, a positive development has taken place in many countries, with the growth of a number of private, quasiprivate, not-for-profit and community-based organisations taking over many of the tasks and services that used to be provided by government.

In order to improve the effectiveness of this institutional complex *vis-à-vis* the rural poor, the public sector should support the institutional strengthening and empowerment of rural communities and farmers' organisations and put in place policies and programmes that foster institutional innovation at all levels.

8. Differentiated policies for agricultural innovation must be designed to fit the needs and possibilities of different situations, such as the three described in this paper. Given the range of situations, it makes little sense to commit to any single option. In general terms, one can recognise three different strategies to develop AKIS in support of poverty reduction:

(a) In situations where agricultural innovation is spurred by favourable contexts and asset positions, and where agriculture is a profitable and competitive enterprise, much of the impetus will be due to the action of market forces, and specifically to the profit opportunities found by commercial farmers and private firms and entrepreneurs in the services and agro-industrial sectors. In this case, public policies and interventions should have the goal of exploiting the intersection between public goods and private interests.

(b) The second set of strategies involves millions of small farmers in developing countries who have the incentives to embark on market-oriented agricultural innovation processes, but lack the capacity to respond fully to this favourable context, either because their assets are too limited, the productivity of such assets is low, or because the transaction costs they face are too high. This group of small family farms represents the best opportunity for linking agricultural innovation and poverty reduction policies in developing countries, and as such they should be the primary target of national and international efforts. Developing the full potential of these small farmers and their communities requires pro-active, marketoriented public policies.

The third set of strategies applies to rural households lacking most types of assets aside from unskilled labour and sometimes a very little land, which at the same time operate in unfavourable environments. Under these conditions, poverty reduction policies and programmes must be even more broad-based than in the second case, and will normally include interventions aimed at trying to move these households towards a better condition, by enhancing their asset position and or improving the context in which they operate. In the absence of such changes, the potential for agriculturebased development will remain very low. Successful innovation in these conditions depends on building local institutions, networks and organisations that help to mobilise the very scarce resources of these communities and link them to external networks.

Chapter 10

Impact of Agricultural Policy Reform on Regional Rural Poverty

The direct income support policy, by which producer price subsidies in agriculture were replaced, targeted to mitigate potential short-term adverse impacts of the subsidy removal, besides to facilitate the transition to e¢ cient production patterns. The main objective in this study is to analyze the impact of this policy on rural poverty. Intra-regional di¤erentiation of the factors that a¤ect rural poverty is given a particular emphasis. The model predicts that the direct income support policy has an insigni.cant impact on rural poverty gap. We also conclude that the main result does not change for di¤erent agricultural regions.

Introduction

In Turkey, accordingly with the disin.ation policy embarked in early 2000, an agriculture policy reform was also in place to achieve the goal of .scal stabilization program. Structural reforms addressing the elimination of producer price subsidies in agriculture and replacing them with a targeted direct income transfer program within limited time frame were introduced. It was aimed by the direct income support policy to mitigate potential short-term adverse impacts of subsidy removal and at the same time to facilitate the transition to e¢

cient production patterns. If the agricultural policy reform, which targeted to eliminate income instability, did really achieve what was expected from this policy, then there should have been a positive impact on income distribution and a decline in number of people, who were living below poverty line. Therefore, the main objective in this paper is to analyze the impact of agricultural reform on rural poverty. Intra-regional di□erentiation of the factors that a□ect rural poverty is given a particular emphasis such that these factors are investigated whether to differentiate between various agricultural regions.

In the first part of the study a poverty analysis is conducted. The quantitative household survey data is used in the analysis. Firstly, the daily calorie-need of an adult is determined in a monetary approach. setting.1 And then, costs of food and non-food basic requirements of an adult and of each household are calculated for the selected regions of Turkey. Regional rural poverty lines are established based on the information gathered above. Each household.s welfare level is compared with regional and nationwide poverty lines to .gure out who lives in poverty and in food-poverty in rural areas of Turkey. In the second part, an econometric analysis is carried out to explore the determinants of rural poverty in these specific rural areas. By using the cross-section econometrics, rural poverty gap indices are estimated based on the education status, demographic structure, work status, source of income, income level of the households, and on agricultural input use and agricultural structure, as well. Searching for divergences between regions in terms of the factors that a□ect rural poverty and divergences among factors in terms of their influences on di□erent poverty levels are given particular emphases. Finally, besides micro indicators, impacts of macroeconomic indicators are examined for better understanding of the factors behind rural poverty.

The paper is organized as follows: regional and na-tionwide rural food-poverty levels are determined in the next section. In section three, poverty lines are established. In section four, econometric analyses are achieved. Finally, it is concluded in section five

The Model: Regional, Rural, Food-Poverty Lines

The micro-level data used in the analyses is obtained from the World Bank coordinated project completed recently.2 ;3 It consists of social, economic and demographic information covering 5507 rural house-holds chosen in 500 villages with population less than 2000. These vil- lages/households are chosen such that they are involved intensively with the production of 8 major and traditional agricultural products: wheat, tea, hazelnuts, maize, cotton, sugar beet, olives and tobacco.

As the collected data belongs to rural areas with population less than 2000, these households are presumed to re.ect the rural people, whose source of income is mainly agricultural sector. Therefore, by using this data, regional and nationwide, rural food-poverty lines, as well as food-poverty ratios are determined in 4 steps.

Daily calorie requirement for an adult

The required calorie approach is used to calculate food-poverty lines in the rural areas. This approach calculates the daily calorie-intake required for an adult. Baysal calculates the daily calorie required for labor force and di¤erentiates the calorie-requirement by gender and age groups. In this study, working age groups are the focus. For each age group, the required calorie-intake is calculated as the average calorie requirement of males and females. An adult.s daily calorie-requirement is determined by taking the average of various age groups. calorie needs. Based on this methodology, an adults daily calorie requirement is calculated as 2450 units.

Food equivalent of the required calorie

The food equivalent of the calculated daily calorie-requirement is determined according to the food-basket prepared previously by Dumanl. Duman has prepared the food-basket that provides the daily calorie-intake required for one labor. In Duman's work, the amount of daily required calorie is 3500 units due to his focus group being heavy-working labor. Therefore, the amounts of speci.c food items appeared in this basket is multiplied by 0.7, and the consumption amounts are adjusted as to provide 2450 calories per day. 2.3 Monetary value of the required calorie.

The next step is to .nd the monetary value of the required calorie in each region. Consumer prices of food items included in the basket are used to calculate the monetary value. Household based food-price data is obtained from the household survey. By using this data, average price paid per gram of each food is calculated for all selected regions, and for the country as well. Each region.s own average prices are considered, while calculating separately the value of food-basket in each selected region. Furthermore, average prices obtained from the whole sample are used to calculate the country-level monetary value of the basket. Region-level monetary values of the basket providing the aforementioned daily calorie required for an adult are de.ned as the region-level food-poverty lines. Similarly, the country-level monetary value of the basket is de.ned as the nationwide food-poverty line. It is worth mentioning that these food-poverty lines are de.ned as daily bases. Therefore, daily food- poverty lines are multiplied by 30 as to give monthly food-poverty lines.

Food-poverty ratios

Food-poverty ratios are determined by comparing households.welfare levels with food poverty lines. In the existing literature, either households.income levels or their level of expenditures is taken into ac- count to proxy households.welfare. Households.food and beverage expenditures are considered, in this regard. Alternatively, food poverty lines could be com-

pared with households.total expenditures. However,

if this was the case, households who were above the food

poverty line with respect to their total expenditures could still be below this line, when their food and beverage expenditures were considered. Therefore, instead of the total expenditures, the expenditures on food and beverages are compared with the food-poverty lines in order not to underestimate food-poverty ratios. It is worth mentioning that food poverty lines are calculated for one person. Hence, food-poverty lines have to be converted into their household equivalents. For this reason, the "adult equivalent scale" is taken into account to calculate the adult equivalents of people of various ages in households. Households. monthly required expenditures are divided by adult equivalent number of people in each household, and hence per adult food and beverage expenditures are calculated.

Poverty Ratio by Regions

When the region-level food-poverty lines are considered, the highest and lowest lines are observed in the third and sixth agricultural regions, respectively. This fact may tell that the highest and lowest food prices are in the third and sixth regions, respectively. Furthermore, the food-poverty lines are below the country average in the .rst, fourth, .fth, sixth and eight regions, whereas they are above in the second, third and seventh agricultural regions. When the ratios of people under the food-poverty lines are considered, the lowest and highest rates are observed in the eight and .rst agricultural regions, respectively. turn related to the consumption of goods and services.

Regional and nationwide poverty lines are determined by adding non-food expenditure to the food- poverty lines. Firstly, all households in the sample are grouped into 1% ranges according to their adult equivalent monthly food expenditures. Secondly, the groups that capture the food-poverty lines are labelled. And then, the average non-food expenditures of households in these labelled groups are calculated. Finally, the calculated average non-food expenditures are added to corresponding food poverty lines to reach region-level and nationwide poverty lines. After that, these poverty lines are compared with total expenditures. Note that, total expenditures are considered as the means of comparison, as the rural poverty lines are determined by taking both food and non-food expenditures into account.

It is worth mentioning that a region.s relatively high poverty line can be due to either a high food- poverty line or high non-food expenditures in that particular region. Besides, relatively high regional prices of both food and non-food items are also effective factors. As shown the highest poverty line is observed in the .rst agricultural region, while the lowest is in the sixth region. The food-poverty line in the .rst region is below the country average, however, its poverty line is well above. As might be expected, the highest poverty ratio is observed in the first region. Apparently, this ratio pulls the countrys average upwards.

Estimating Rural Poverty Gap

In this section, the factors that a¤ect rural poverty gap are estimated by employing cross-section econometric models. Poverty gap statistic of poor house-holds is considered as the dependent variable. In the previous section, 1176 of 5507 households are found to live in poverty. The poverty gap statistic of these households is calculated by taking the difference between the monthly adult equivalent poverty line and per adult monthly total expenditures. Factors that may a¤ect poverty gap are determined according to theoretical approaches and empirical findings of the existing literature on the poverty problem in Turkey. Indicators related to education level, demographic structure, working conditions, source of income, level of income, agricultural input use and agricultural structure are found to be the potential factors that may have impact on poverty and poverty gap. Explanatory variables related to above indicators are derived from the aforementioned survey.

Total sample of 1176 poor households is checked for possible outliers. Five households are excluded, in this regard. The multi-collinearity problem is also checked by using correlation matrix of the explanatory variables and no serious collinearity is detected. Another important problem might be heteroscadasticity, and thus, White.s heteroscadasticity consistent estimators are used.

Econometric analyses are completed in four stages. A benchmark model is constructed and estimated in the .rst-stage. The implicit form of this model with expected coe¢ cient signs is given in equation. In the second stage, the focus is on .nding whether the explanatory variables have varying impacts on poverty gap in different regions. In the third stage, poverty gap variable is tested as to reveal whether certain thresholds are embodied. And .nally, impacts of macroeconomic and nation-wide indicators, which are emphasized in the international applied literature, are tested.

The two intercept dummies regarding the years of education hh9e and hh58e are both expected to have decreasing impact on poverty gap. The other intercept dummy hmlit representing illiteracy is, however, expected to have increasing impact. Similarly, as the number of household members increases, a rise in poverty gap is expected. Four variables are included with regard to the household income. Among these variables, wages as a share of non-agricultural income salaries income as a share of non- agricultural income and agricultural income as a share of non-agricultural income are expected to decrease poverty gap. However, the share of direct income payments in agricultural income may decrease or increase poverty gap. Capital accumulation in rural areas is represented by three variables: total number of sheep total number of cattle and total sown area (sa). The sheep, cattle and sown area can be considered as both .nal and investment goods. Therefore, a rise in these variables is expected to have decreasing impact on poverty gap. Additionally, an increase in agricultural input use per agricultural area is expected to reduce poverty gap, as an increase in yield is expected due to an expected increase in production since the input use increases. Similarly, a rise in the share of irrigated land is also expected to increase the yield and income, and hence to reduce poverty gap. A rise in number of secondary products and in income can also be anticipated, as the agricultural yield increases. Therefore, a rise in number of secondary crops would have decreasing impact on poverty gap.

To proxy the distribution of agricultural land own- ership, agricultural land per household members is considered. An increase in this variable may be interpreted as the increase in production scale, and hence an increase in per capita income. Poverty gap is anticipated to decrease by the increase in per capita owned agricultural land. In Turkey, prices of annual crops seem to be more volatile compared to prices of perennial crops and trees. Modeling the impact of this price volatility on income and poverty gap, especially in di¤erent regions, becomes important due to the regions.varying crop patterns. Therefore, the relationship between the type of agricultural production and the poverty gap is analyzed by employing three variables the share of annual and perennial crop areas and of poplar tree area in total agricultural land. As the share of perennial crops and of poplar trees increase, a decrease in poverty gap is expected. The opposite impact is anticipated for the annual crops.

The first-stage estimation

The base model, whose implicit form is given in is estimated in this stage. According to the estimation results presented in poverty gap seems mostly autonomous. The independent variables are able to explain only 16 percent of the variation in poverty gap. But, majority of the coe¢ cients have the anticipated signs, however their statistical signi.cances show variations. It is worth noting that the variation in an economic indicator is explained by both economic and social factors in the current model.s setting.

Therefore, elasticities are not calculated. For the same reason, some of the variables having the expected coe¢ cient signs, but being statistically insigni.cant, are not excluded from the equation.

Base Model Estimation Results

Estimation results reveal that both number of people in a household and education level of a household head a□ect poverty gap. We find that poverty gap is increasing by a rise in the former, while it is decreasing by a rise in the latter. Furthermore, we find that holding of a household-head particularly a secondary- school degree is quite in.uential on poverty gap. Similarly, if all household members are illiterate, this has an increasing impact on poverty gap. In fact, the autonomous part of poverty gap increases with illiteracy. As might be expected, more years of education imply less autonomous poverty gap. Moreover, we find negative relations between poverty gap and the indicators relating to household income wherein these indicators have statistically signi.cant coefficients. Especially, share of non-agricultural income is quite in.uential on poverty gap. This result is quite expected as the agricultural income is more unstable compared to wages salaries and non-agricultural income.

Share of direct income payment in agricultural income is observed to have insigni.cant impact on poverty gap. As it is not known how much of these payments are spent on agricultural input use or on mechanization to have an increase in yield and an increase in income, and as these payments are paid for sown area, but not for the amount produced or for the amount of yield, the direct income support is not actually leading farmers, but instead price movements do.

Capital stock of rural families seem to have decreasing impact an increase in capital stock means a rise in potential income on poverty gap, but neither of the regarding variables has statistical signi.cance. Regarding the agricultural structure, a rise in agricultural input use per hectare and a rise in share of irrigated land are both expected to increase the yield, and hence increase agricultural income. Thus, these are expected to lower the poverty gap. Estimation results are in line with the expectations, and the results are statistically signi.cant. Similarly, a rise in number of secondary crops has the same impact. Alternative use of agricultural land between annual crops, perennial crops and trees are found to have expected signs, but they are statistically insigni.cant. The last indicator regarding the agricultural structure is per capita owned agricultural land in a household. In fact, this variable is used to proxy the distribution of agricultural land. In line with the anticipation, a rise in this variable is found to decrease poverty gap and to be statistically signi.cant.

The second-stage estimation

In the second-stage estimation, the varying size of the autonomous poverty gap and of impacts of explanatory variables in various regions are evaluated. Firstly, the base model is extended as to include regional intercept dummies. It is statistically found that the autonomous poverty gap is signi.cantly different from the average only in .ve regions. Secondly, based on the model above, slope and interaction dummies are introduced in the regions, where the autonomous poverty gap is statistically different from the average. In these two estimations, the statistically signi.cant variables of the base model are also found to be signi.cant and to have expected effects showing the consistency of the base model. However, the models in general explain between 17-23 percent of the variation in poverty gap. Moreover, regional intercept, slope or interaction dummies are found to have only slight impacts.

The third-stage estimation

The poor performance of the base model and its extended versions in terms of the adjusted R2-statistic bring to mind the question of whether there is a threshold level of poverty gap that the base model can explain better below or above that threshold level. In this regard, poverty gap data is regrouped into five. To see impacts of the base model.s independent variables on different poverty gap groups, new intercept-dummy variables are introduced as to reflect these groups. Estimation results reveal that the autonomous poverty gap does change signi.cantly within the poverty gap groups. The percentage variation of autonomous part from constant term of the equation in each poverty gap group is given in

Dummy Variables

Another .nding is that in the upper two groups, or in the two groups just below the poverty line, a lower autonomous part and a higher explanatory power might be expected. However, the regressions that model the impact of base model.s indicators on poverty gap statistics of these two groups do not result in a signi.cant improvement.

The fourth-stage estimation

Household-level data has not provided enough evidence so far in explaining the factors that have significant e¤ect on poverty gap. Therefore, it is fair to say, up to this point, the higher percentage of poverty gap remains autonomous in both various regions and various poverty gap groups. In this section, we mainly test the hypothesis of whether the macroeconomic and nationwide indicators factors are also ineffective in explaining the poverty gap. The existing applied literature in this regard mostly focuses on linkages between growth, income distribution and poverty. However, their findings do vary. In order to test the above hypothesis, the database is rearranged as to provide province based averages. After that, new province level varables are introduced to the model such as public agricultural investments as a share of total public investment , growth rate, marketed value as a share of total production value urbanization rate and level of socioeconomic development.

Explanatory Factors

When macroeconomic and nationwide indicators are introduced to the model, the overall explanatory power increases to 47 percent. This result is at least an explanation for the autonomous part of poverty gap, which cannot be reduced by the previous econometric analyses in this study. In the current model, poverty gap decreases by an increase in either household based average number of years in education, number of secondary products, per-capita owned agricultural land or share of owned land in total, whereas it increases by a rise in either share of input expenditures in agricultural income or share of direct income support in agricultural income. Furthermore, among the macroeconomic and socioeconomic variables, province based growth rate, urbanization rate and socioeconomic development level have insigni.cant coe¢ cients. Finally, we find that both public agricultural investment as a share of total public investment and marketed value of production as a share of total value have the expected coe¢ cient signs. Poverty gap decreases by an increase in these variables.

Conclusion

Preliminary outcomes provide the information that certain income groups might be suffering from structural poverty, which means these people cannot change the socioeconomic conditions that they are living in, and even if they experience a change in these conditions, these changes do not have signi.cant effect on their level of poverty. In addition, household based, socioeconomic indicators have very limited power in explaining the households. poverty gap, particularly for the most poor ones, though we .nd macro oriented policies more influential, especially in breaking the structural poverty.

Chapter 11

Land, Interlinking Markets and Rural Poverty in Transition Countries

This paper analyses how various factors have affected rural poverty in transition countries of Central and Eastern Europe and the Former Soviet Union. The paper analyses first how land reforms and land markets have affected rural poverty, and second how non-land factors have played a role, including the interlinking of input and output markets.

Introduction

Much has been written about the transition process that occurred in former centrally planned economies of Central and Eastern Europe and the Former Soviet Union. A large literature exists on the changes and transformation of the agricultural sector and its consequences for output and productivity.

Much less was known until recently on how all these changes have affected welfare, and in particular poverty, in rural areas. While development scholars and policy makers have long had a strong interest in poverty differences between rural and urban areas, information on transition countries was mostly limited to anecdotal evidence and a few country-specific analyse.

Two recent studies for the first time provide cross-country comparative evidence on rural poverty in transition. The first section of this paper will draw importantly on Macours and Swinnen. We present an overview of poverty in rural areas in transition countries, compare it to urban poverty, discuss how rural poverty has changed over the past years, and provide hypotheses to explain differences across countries and changes over time.

Land reform, policies, and markets had an important impact on rural poverty. While much has been written on land reform, much less is known on the developments of land markets. In this paper we will first identify some general mechanisms through which land reform and markets affected rural poverty. We then use some new evidence on the development of land markets to add insights on this process. For the latter, we draw on a recent study by Swinnen and Vranken.

An important development in transition countries is the interlinking of markets. This is having important impacts on efficiency and equity. These developments are summarized in section 4. The final section presents several distinct ways of how rural poverty is affected by land reform, markets and interlinking.

This gap in our understanding and research is particularly striking for two reasons. First, a large share of the population in transition countries lives and works in rural areas and the share is largest in the poorest countries. For example, in Central Asia and part of the Balkan countries, more than 50% of the people live in rural areas. In all transition countries more than 25% of the population is rural. An even larger share of the poor live in rural areas. Rural areas in the region, as in most of the rest of the world, have a disproportionate share of poor households. Second, the lack of attention to rural poverty is most striking in those countries where transition caused an increase in rural poverty, and often dramatically so.

Rural Poverty in Transition Countries Data

This paper relies on new comparable data of poverty in transition countries that were calculated by the World Bank based on household level surveys. In particular, we rely on the head count indices which are based on a common absolute poverty line of 2.15$ a day in 2000 PPP for all countries and years. Adjustments for within-country price differences allow comparing rural and urban areas.

For both conceptual and practical reasons, we split up our analysis in pre-1998 and a post-1998 period. The pre-1998 period was characterized by the fundamental economic changes related to the move from plan to market, and the massive unanticipated economic collapse in which it resulted. Trends were reversed after 1998, as the Ruble crisis and the economic recovery after deflation in Russia, as well as the maturation of the market system and the incentives related to EU accession, proved to provide a strong impetus for economic growth in the region. Also for most countries the most drastic changes that affected rural areas, liberalization of agricultural input and output prices, land privatization and the restructuring of large-scale state and collective farms, occurred in the pre-1998 period.

Rural poverty and urban-rural differences

There is an enormous variation in rural poverty across the region, with rural headcounts ranging from 0.3% in Hungary to 77% in Kyrgyzstan in 2002. Absolute poverty levels largely correspond to differences in overall income levels between countries. Poverty is high in Central Asia and the Trans-Caucasus, and it is the lowest in Central Europe.

Income poverty in rural areas is significantly higher than in urban areas in the vast majority of transition countries. Rural versus urban poverty indicators from other sources, be it poverty incidence ratios (based on relative shares in the poorest quintile) or poverty headcount ratios using national poverty lines also show that for the vast majority of transition countries incomes in rural areas are much lower than in urban areas and rural poverty is much higher than urban poverty.

Intriguingly, several of the countries with the most extreme differences are Eastern European EU accession countries. Romania, Latvia, Lithuania, and Bulgaria have the highest differences between rural and urban poverty risks. The differences are also very high for Russia and Kazakhstan. In those countries the ratio of rural on urban headcount is higher than 1.5, meaning that poverty risk is more than 50% higher in rural than in urban areas.4 In contrast, poverty is significantly lower in rural areas in Belarus and in two Trans-Caucasian countries, and this finding is consistent across years and indicators.

Non-income poverty indicators help shed further light on rural welfare and rural-urban differences. Infant mortality in rural areas is highest in Bulgaria, Romania, Russia, and in Central Asia . In all these countries, the infant mortality rate is more than 15 deaths per 1000 live births in 2002. Infant mortality is lower in the Trans-Caucasus, and the richer Eastern European countries. Looking at rural-urban differences we find a more mixed picture than for income poverty. Infant mortality in rural areas is about the same as in urban areas in 30% of the countries, higher in 40% and lower in the remaining 30%. Several Eastern European EU accession countries such as Bulgaria, Romania, Latvia and Lithuania have much higher rural non-income poverty, relative to urban. Another country with much higher rural indicators is Belarus. In Central Asia, infant mortality rates are substantially lower in rural areas. Also in Estonia, and in Armenia and Georgia, rural indicators are better than urban.5

Changes over time

Summarizes the available data on changes in income poverty changes in rural and urban areas over the 1998-2002 period for the countries for which time-series data is available for the same years, i.e. Moldova, Georgia, Romania, Russia, Belarus, Lithuania, Hungary and Poland presents the same graph by country. These figures suggest some interesting patterns and conclusions.

- The developments are remarkably similar across

countries. In general, poverty initially went up almost everywhere, and a decline in poverty started in 1999 or 2000. Infant mortality in rural areas declined in virtually all ECA countries between 1998 and 2002. The declines were rather similar across countries, on average between 15% and 30 %.

- Rural and urban poverty have moved in strong parallel over this period. This suggests that the type of growth which has occurred since 1998 has also been good for rural poverty reduction.
- There continues to be more poverty in rural areas. Rural poverty is declining (almost) as fast as urban poverty, but not faster. Hence, the key question is not "why rural poverty is lagging behind ?" but rather, "why is it not catching up ?".

In this paper we forward a series of hypotheses to explain these patterns, and the country differences in rural poverty levels. Given the parallel developments of rural-urban poverty in recent years, the levels of poverty and rural-urban differences in 1998 are important to understand.

Effects of Land and Price Reforms on Rural Poverty Developments The role of agriculture

Agricultural reforms are key to understand rural poverty for several reasons. First, agriculture is a major income source and employs a large share of the rural population in most transition countries, and especially so in the poorest countries. For example, agriculture accounts for more than 40% of rural income and employs 40% or more of the *total* work force – and a much higher share of the rural workforce in several countries in South Eastern Europe, the Caucasus and Central Asia. These are countries with high rural poverty rates. Second, in the countries where the share of agriculture in employment is (much) lower, the poorest are usually either directly or indirectly linked to agriculture, either because they are involved in subsistence farming and or because they have been laid off by large farms and have not been able to find alternative employment. In some countries agriculture has also absorbed many workers laid off in other sectors. Third, collective and state farms were historically the suppliers of rural services, and continue to do so in some countries. Access to rural services is related to the restructuring of the farms.

Reforms in agriculture include several factors, such as price and subsidy policy reforms, trade policy, land reforms, farm restructuring, privatization of agri-food companies and liberalization of markets. These reforms have affected rural households in several ways.

Price and trade liberalization

The prices and trade liberalization and subsidy cuts caused a dramatic decline in farm profitability and rural incomes. Agricultural production and food consumption were heavily subsidized under the communist system. Macro-economic reforms coincided with price liberalization and subsidy cuts in the early years of transition. The result was major reductions in the support to agriculture. Reduced domestic demand with falling incomes and subsidy cuts were reinforced by falling foreign demand and increased import competition with trade liberalization. Agricultural terms of trade declined dramatically, i.e. between 40% and 80%, causing a large part of the initial output decline. It most likely leads to an 6 increase in rural poverty. Belarus, where such reforms have not been implemented and very extensive subsides of agriculture are continuing, provides an interesting benchmark in this regard. It stands out for having low levels of rural poverty and small rural-urban differences.

The recent improvements in rural poverty in CIS countries are likely due to an improvement in agricultural prices and overall liquidity and growth in the economy, which have pushed up wages and improved services and wage payments. First, in several of the large CIS countries the 1998 financial crisis had an important impact on agriculture and the rural population. After the negative shock of the crisis, the associated devaluation caused an important improvement of the competitiveness and the profitability of domestic agricultural producers who saw relative prices rise substantially. For example, in Russia, the grain over fertilizer price ratio, which had fallen by around 80% in the early 1990s, increased strongly after 1998. Real agricultural prices increased more than 20% in Belarus, Russia, Kazakhstan, and Ukraine. After reaching their lowest levels in 1998, output and yields rebounded. Relative prices in several Eastern European countries declined in the same period.

Land reforms

Land reforms and farm restructuring initially caused important disruptions and sometimes reinforced output declines and income falls. However, later these reforms have been sources of growth and poverty reductions.

There are however major differences between the countries in how land reform, farm restructuring and the development of land markets have affected rural development and poverty. These differences can be explained by a combination of initial conditions and reform policies. The effect of land privatization and the extent of farm restructuring and resulting farm sizes was in part determined by the capital intensity of the prereform production process, and the choice of the privatization method. While land was restituted in most of Eastern Europe, land was distributed in shares in many of the large CIS countries, and distributed in physical plots in some low-income countries in Central Asia, the Trans-Caucasus and Albania.

Differences between urban and rural poverty in some of these low-income transition countries are smaller in part because land distribution to poor rural households with labor-intensive agricultural technology (a) induced important productivity gains, and (b) enabled selfemployment in agriculture, which helped mitigate the negative shock of transition. To illustrate the importance of this argument it is helpful to consider these countries as part of a broader pattern.

shows the remarkable correlation between labor intensity in agriculture and the growth of household farming. In labor-intensive rural economies, access to land through distribution of land plots to rural households induced important productivity, and income gains in rural areas. This is what happened in Albania in the early 1990s, in Kyrgyz Republic in the mid 1990s, in Azerbaijan in the late 1990s, and probably in Moldova after 1999 which may explain why Moldova is the only country where rural poverty declined much stronger than urban poverty over the 1998-2003 period.

Interestingly, the distribution of land in the low-income labor-intensive countries not only led to a shift to individual farming and an increase in yields in particular those of labor intensive crops but also to a large inflow of labor into the agricultural sector. This suggests that households could improve their welfare by shifting to agricultural self-employment after the land distribution. As a result average labor productivity declined. However, value added in agriculture per worker started increasing after 1996 in the countries that shifted to individual farms in the Trans-Caucasus and Central Asia. This is consistent with early evidence of positive total factor productivity effects of the shift to individual farms in Central and Eastern Europe.

In many higher income countries in Central and Eastern Europe, and in middle-income countries, such as Russia and Kazakhstan, individual farms did not become dominant however, and large farms have remained important in several areas. Moreover, even in countries where individual farms became dominant after restitution i.e. the poorer EU accession countries, these were often limited to semi-subsistence farms with limited productivity gains. Both outcomes can be explained by the more capital-intensive nature of agriculture in these countries, which would require access to other inputs, including credit, and commercial markets in order to obtain a comparable income as in the rest of the economy. In Central Europe, land privatization through restitution introduced hard-budget constraints for the large-scale farms, that underwent major restructuring and a large outflow of excess labor out of agriculture. This was facilitated by the availability of social transfers which mitigated the effects on rural poverty.

In contrast, in the middle-income CIS countries, distribution of shares and lack of alternative income sources, led to the initial continuation of excess labor on large-scale farms. The restructuring of these large-scale farms did not lead to significant income gains in rural areas. Indeed, differences between rural and urban poverty rates are large, and rural poverty reduction in these countries needs to rely on finding productive use of the underemployed rural labor, by the creation of off-farm employment or the reduction in labor mobility.

The Development of Land Markets and Household Incomes

Land markets are active, relatively transparent, and well developed in some transition countries while embryonic, informal or opaque in others. In general, households and farms in Central Europe have much less difficulties in accessing land than in some of the CIS countries. These variations reflect the different stages of reforms and economic development, and the regional environments these countries operate in. For example, in several new EU member states land prices and market activities have grown rapidly with the combination of increased demand due to higher subsidies, prices and productivity, the inflow of foreign investment, and new land legislation. The development is slower in most countries further east.

The main form of land exchange in transition is through renting of land. Buying and selling of land is more difficult than renting everywhere. However, the share of land rented in total land used varies enormously: from more than 90% (e.g. Slovakia) to 10% or less. The importance of land renting is strongly correlated with the importance of corporate farms in land use. Corporate farms rent most of their land. Larger family farms operate on both owned land and rented land. They have typically enlarged their farm by both purchasing and renting land. Small family farms and subsistence farms operate mostly on owned land, and rent in small plots of additional land at most. They are more likely to rent out land than to rent in land.

This is reflected in very different land markets across countries. First, in labor-intensive agricultural economies where small-scale family farms dominate (such as in Albania, Azerbaijan, and Kyrgyz Republic), there is relatively little rental going on, all of it household to household and mostly informal. Second, little land renting is also going on in land intensive agricultural economies where land was initially distributed as shares but where large scale corporate farms have been able to acquire ownership of the land. Corporate farms, or larger agro-holdings, acquired vast amounts of land either through bankruptcy proceedings and sales e.g. vertically integrated grain companies in Northern Kazakhstan and agro-holdings in Russia, or through changes in government policy which transferred ownership rights to farms using the land in current Kazakhstan. In these systems a large share of the land is owned by large corporate farms, and very little renting is taking place. Third, in contrast, in capital intensive agricultural economies in Central Europe where large scale corporate farms dominate but land ownership was restituted to households (such as in Slovakia, the Czech Republic, and parts of Hungary and Bulgaria), there is very extensive renting of land going on, mostly from households to large scale corporate farms, often based on formal contracts. Land rental markets are well developed and prices increasing, especially in those countries which have acceded to the EU.

Rental contracts differ in length, formality, type of payment, etc. both across countries and across farm types. In general, corporate farms tend to have longer contracts, more formal contracts, and are more likely to pay in kind and to pay lower prices than family farms. Renting to family farms is more likely to be paid in cash, with more informal contracts and for shorter contract lengths.

Effects on Equity and Efficiency

Micro-empirical evidence indicates that the effect of land markets on incomes, poverty, and efficiency is conditional and depends in part on the players involved in the market. Land renting between households has a positive effect on equity and efficiency. First, households with more human capital access land through a combination of buying and renting land, and rental markets contribute to increased returns to labor on family farms. Second, older and less educated households rent out land to get additional incomes, and those who can rent out their land if they wish, have higher welfare. Third, rental markets reduce inequality of access to land by transferring land from households with high land endowments to those with low land endowments. In contrast, sales markets seem to contribute to inequality of land ownership. Fourth, larger family farms combine renting and buying of land to enlarge their farm operation. However, where corporate farms dominate the land rental market these effects can be very different. While corporate farms may be efficient farming organizations in some regions and for some farming activities, transaction costs and regional monopoly power of corporate farms in the land market are causing negative equity and efficiency effects in several countries. In several countries, corporate farms are using more land pay lower rental prices than family farms, are more likely to pay rents in kind than family farms (who pay cash), have rental contracts of longer duration (locking in land), and often use their political powers/relationships to influence policies that shift effective land property rights in their favor.

The problems are more serious (a) where land is owned as shares than where households have physical plots; (b) where land is initially allocated in the middle of large consolidated plots, (c) where the costs of withdrawal is expensive, either because there is considerably uncertainty on the (co-) owners of the land or because ownership is highly fragmented through the combination of restitution and an egalitarian historical rural land ownership structure (eg Bulgaria and Slovakia) or because the registration costs are high and or corporate farms consider the land as quasi-property of the farm.

Government policies may not directly favor corporate farms, but the implementation may be biased towards corporate farm interests, because of technical requirements related to land exchanges and withdrawal procedures, because complex and expensive land registration procedures, and because of established relations between farm and officials. In extreme cases, such as Kazakhstan, government policies have reallocated land rights from (small) owners to (large) farms.

Other Constraints

Access to capital is an important constraint for many farms in the land market, not only for buying land but also for renting land. The accession to the EU with increased demand, prices, productivity and subsidies has seriously reduced this constrained in the new EU member states and stimulated land renting and prices. The financial constraints of farms remain important constraints in less developed transition countries, including in South and Eastern Europe. Problems with property rights imperfections may remain after full titling and after cadastres and information systems have been introduced. One example is the existence of co-ownership of land in e.g. Bulgaria and Slovakia. This causes significant transactions costs in land exchange and contributes to low rental prices for owners.

Imperfections in input and output markets and interlinking

The privatization and restructuring of the agri-food chain caused major disruptions in input and output markets. Widespread contracting problems were long payment delays or non-payments for delivered products, causing drains on cash flow, and constraints in accessing inputs and selling products. At the same time, food-processing companies have problems obtaining quality supplies. The problems are worsened by the lack of public institutions necessary to support market-based transactions, such as for enforcing property rights and contracts. Private contractual initiatives have emerged to overcome disruptions of supply and poor public institutions for governing exchange of inputs and output markets. This occurs through interlinking markets and vertical coordination in agri-food supply chains, which is an important and growing phenomenon in transition countries. Traders, agribusinesses and food companies have started contracting with farms providing inputs and assistance in return for guaranteed and quality supplies. Successful vertical contracting typically includes conditions for product delivery, prompt payments, as well as providing farm assistance programs for suppliers. Farm assistance can include input supply programs, investment assistance, trade credit, bank loan guarantees, extension and management advisory services, etc.

Enforcement of these contracts is problematic as public enforcement institutions are absent. Companies try to create "self-enforcing contracts" by designing the terms of the contracts such that nobody has an incentive to breach the contract. They try to enforce contracts by interlinking markets. The enforcement of the credit transaction occurs through the output market.

Successful vertical coordination has important positive effects, both direct and indirect. The direct impact is on increased output and productivity of the processing company that initiates vertical contracting. Indirectly, contract support measures have positive effects on farm productivity and product quality. Measures with the greatest impact on yields were specialist storage cooling equipment in dairy, veterinary support and physical inputs. Prompt payments, guaranteed prices, and market access also had large positive effects. Quality of output improved strongly in response to specific programs. Direct loans and loan guarantee programs stimulated farm investments. Programs that assist farms in accessing inputs mainly feed enhance investment indirectly by lowering input costs, or reducing transaction costs in accessing inputs, improving profitability.

An important concern is that interlinking markets and vertical coordination (a) exclude many farmers, in particular small farmers, and (b) may lead to rent extraction. First, transaction costs favor larger farms in supply chains. Second, small farms are more constrained for making necessary investments. Third, small farms typically require more assistance per unit of output.

Therefore, companies may prefer working with relatively fewer, larger, and more modern suppliers. In reality, we find that, so far, companies work with surprisingly large numbers of suppliers and of surprisingly small size. There are several reasons. Companies may have no choice if small farmers represent most of the supply base. Contract enforcement may be more problematic with larger farms. Farms' willingness to learn and attitude are more important than size in farmprocessor relationships. Small farms may have cost advantages in labor-intensive production activities. Processors may prefer a mix of suppliers. Cooperatives are more likely to work with small farms than corporate companies, either domestic or foreign. However, it may also be that, so far, the transition characteristics give suppliers of quality produce a strong bargaining position in the chain as there is an excess demand for quality produce.

Other Causes of Rural Poverty Developments

The rest of the economy

Obviously, rural poverty also depends on what happens in the non-agricultural sectors of the economy. Rural poverty was affected by general developments in several ways. First, in several cases a broader economic crisis triggered changes and radical policy reforms – as had happened in China in 1978. This was the case in Albania in 1990 when food shortages triggered rural revolts and spontaneous de-collectivization, in Armenia and Georgia in the early 1990s following natural disasters and war, and in Azerbaijan in 1996 after a major political change. Second, household coping strategies in response to negative transition shocks, in combination with new access to land resulting from these land reforms, resulted in urban to rural migration flows in some countries. With industrial decline, deteriorating urban economic conditions and food insecurity, the best option available for many unemployed urban households was to move to the rural areas. These different developments clearly suggest that urban and rural poverty are interdependent.

Third, more recently declines in rural poverty have moved in parallel with urban poverty decline and overall growth. This can partly be explained by the fact that social transfers are an important part of rural households income and depend on the overall economic situation. Social transfers have increased in recent years, especially in countries such as Russia and Kazakhstan, which have benefited from strong growth in incomes and improvement in the government's fiscal situation. As a consequence, pensions, other social payments, and public sector wages have been paid (on time). Moreover, transfers to hospitals, schools, and other organizations have improved.

The increase in social transfers helped to reduce poverty in both rural and urban areas. That said, there are very important differences among rural households in the extent to which they are affected by economic decline or can benefit from an upturn in economic growth. This will in part depend on their asset and human capital endowments. Land ownership may help them survive through the worst moments, and can also provide additional income when agricultural economy grows, e.g. through renting out the land. Human capital of the household and labor market constraints will certainly affect the households' abilities to participate in growing off-farm opportunities, either through migration or through off-farm rural employment.

Human capital, demographics and migration

The human capital endowments in the rural areas are however drastically affected both by the responses to the initial reforms, and by more recent developments. The reforms affected the demographic structure of rural and urban areas in strikingly different ways across countries. While access-to-land, e.g. through households plots, provided a short-term coping strategy in countries like Russia, the urban-to-rural labor reallocation seems to be of a more permanent nature in the poorer countries of Central Asia and the Trans-Caucasus. The urban-to-rural migration in low-income transition countries can be explained by the land privatization policies that physically distributed land, which helped offset the negative shock to the urban economies.

Labor-intensive agricultural (self-) employment provided an alternative income mechanism in these low-income countries and some urban residents including the young and dynamic ones might have opted for farming on land from rural family members. In countries where land was restituted to the former owners, urban-to-rural migration was more limited and concentrated among the elder population. At the same time, many young people left the rural areas in search of better opportunities. The restitution process, by allocating land to the older population hence had drastic effects on the demographic structure of the rural population in countries like Bulgaria, Romania, Latvia and Lithuania. Land access did not necessarily play the same role in the richer Central European countries where formal security systems, and in particularly pensions, helped people cope with the negative shocks from transition.

In those countries where individuals with higher human capital have migrated from the rural areas to urban areas, the rural areas are disproportionately populated with older and less educated individuals, which exacerbates rural-urban poverty differences.

Indeed, the share of elderly is considerably higher in rural than in urban areas in a number of countries in the European CIS (Belarus, Moldova, Russia and Ukraine) and in Bulgaria, Lithuania, Romania. On the other hand, the proportion of adults between 25-39 is higher in the urban areas in these countries. While studies for some countries have shown that at the level of the household, the presence of a pensioner in the family might actually decrease poverty risk (because of pension income), the unfavorable age distribution is likely to have the opposite effect on aggregate rural growth and poverty reduction.

The outflow of the more educated labor force has affected another important determinant of poverty in the region. Indeed, educational levels differ substantially between urban and rural areas. The relatively low endowment of human capital in the rural areas helps explain part of the urban-rural poverty differences. Inadequate human capital is a very important constraint, not only for agricultural restructuring, but more generally for business development and economic activities in rural areas. Education is positively correlated with enterprise development, both farming and non-farming, and the likelihood of finding another job in the service sector or in industry. Studies find a non-linear relationship between human capital and farming activities. Farm entrepreneurship increases with education, beyond a certain level of education individuals tend to leave agriculture and choose for non-agricultural employment.

The positive impact of education on household welfare is probably due to a combination of factors. More highly educated household heads tend to be better farm managers. At the same time, higher education is associated with more off-farm employment opportunities and higher wages. In addition, it is likely that better educated people were better positioned to benefit from the reform process through their former position, possibly as part of the former farm management, which was associated with better understanding of the reform process and its opportunities, lower transaction costs in accessing assets, a network of important relationships, etc.

Finally, the overall poverty effect of the human capital disadvantages of the rural areas depends in part on the level of private transfers. At the level of the household, labor outflows might help to increase household welfare through remittances. Judging from data on remittances, international migration is particularly important as an income source in Albania, Moldova, Bosnia&Herzegovina, and Georgia. Interestingly, remittances have increased substantially in many other countries in recent years. However, net private transfers tend to be lower in rural than in urban areas. The available evidence suggests that inflow of remittances and other types of private transfers is often lower for rural than for urban areas

Access to services and infrastructure

Access to services and infrastructure in rural areas is worse than in urban areas in all countries, and in many countries much worse. Rural services are relatively better in Central Europe than in the other transition countries. Rural services are particularly poor in South East European countries such as Romania, Albania, and Moldova, and in Azerbaijan and Central Asia. Not only are there less services available, the quality is often lower in rural areas: good doctors, hospitals, schools etc are in urban areas. High quality service providers have often migrated from the rural areas to the cities or abroad.

With transition, accessing services has become more difficult as their supply has decreased, and their price has risen. In addition, the cost of access has increased because travel has become more costly, both in terms of price and quality and availability of transport. In a very real sense, rural villages have become more "remote". An important related problem in rural areas is poor access to information. Interviews with rural households and farmers point at lack of information as a major problem. It constrains their business decisions e.g., agricultural prices and weakens their negotiation power versus traders and local authorities. Poor information has negatively affected households in the farm restructuring and land reform process. In many countries asymmetric information between former collective farm managers and rural households has allowed the managers to manipulate the reform implementation in their advantage.

Patterns of Rural Reforms & Poverty Interaction

The process of the reforms, the implementation, and the effects on market development, both initially and more recently have varied tremendously between transition countries. To summarize how the various reforms and market developments have affected rural poverty, we can identify several patterns.

Pattern 1: Small-scale agriculture-led broad-based growth in rural areas. In most highly laborintensive poor countries with a large share of the population employed in agriculture, such as Albania, Azerbaijan, Armenia and Kyrgyzstan, governments distributed land to rural households in physical plots. This process coincided with a strong fragmentation of farm structures, effectively rapidly shifting land from large-scale former collective farms to household farms during difficult times in the early 90s. There is relatively little rental going on, all of it household to household and mostly informal. The rental markets are constrained by imperfections in the product market (output marketing) and several input markets, in particular credit and labor markets though the later is partly resolved through migration.

Direct access to land leads to shifts towards labor-intensive crops horticulture and livestock production. Rapid and strong productivity gains followed, leading to increases in household incomes. Figure 6 shows that output and yields grew much stronger in a group of countries that followed this pattern. This path has strong commonalities with the Chinese path of agrarian reform and growth.

The countries with strong fragmentation of farm structures have relatively small (and sometimes positive) differences between urban and rural headcounts. This suggests that land access might have helped mitigate rural poverty. The land-individualization process has typically coincided with inflows and absorption of labor into agriculture. In the three Trans-Caucasian countries distribution of land coincided with a strong and sudden increase of labor in the agricultural sector.

Also in the Kyrgyz Republic, agricultural labor grew strongly.

Pattern 2: Large-scale agriculture led productivity growth with major employment reductions but good social security. An entirely different adjustment pattern has been followed by richer 13 This list of patterns is not conclusive and some countries may not fit in here. The selection of the patterns is based on their relevance for the poverty patterns identified in this paper.

Central European countries such as Czech Republic, Slovakia, and Hungary. In those countries, large-scale privatized farms have laid off surplus workers who have either found jobs in other sectors or become unemployed or gone into early retirement. This process has resulted in a relatively rapid and strong outflow of labor from agriculture. There is very extensive renting of land going on, mostly from households to large-scale corporate farms, often based on formal contracts. Productivity growth came mainly from major gains in labor productivity on large farms, rather than from yield increases that only started as of the mid 1990s. The countries in this group were higher-income countries with relatively good social benefits and pensions, and relatively few people employed in agriculture. These higher-income countries could afford to keep rural poverty - and urban-rural differences - low through social transfers, while at the same time benefiting from the productivity gains.

Productivity gains were further enhanced by the inflow of FDI and investment in the food industry. This investment, through vertically coordinated contracting, induced improved access to inputs, credit, technology, and markets for farms, even for small household suppliers. This process started earlier in Central Europe, where general economic and institutional reforms were advancing faster, than in countries like Romania and Bulgaria.

Pattern 3 : Falling between the cracks in poorer EU accession countries. Rural households in poorer EU accession countries (such as Bulgaria, Romania, Latvia, Lithuania) faced the worst of both situations. Their unemployment and pension payments were lower than in the richer central European countries, and households which were laid off by large farms or where collective farms dissipated had to fall back to farming to complement their income. As a consequence relatively few people initially left farming, and there even was (initially) some inflow of labor into agriculture as people laid off in industry fell back to semi-subsistence agriculture.

However, as these countries had relatively developed and capital-intensive agricultural production systems, productivity gains from shifting to small farms were less than in poor labor intensive agricultural countries, while the costs in terms of losses of scale economies and technology disruptions were larger. With limited access to credit, inputs, and technology and few off-farm employment opportunities in rural areas, many rural households remained stuck in semi-subsistence farming to complement their incomes and with many constraints limiting growth or investments. Moreover, access to health and education services was severely affected by the collapse of the collective farm system.

Hence, the land reform and privatization might in fact have contributed to growing rural poverty as both the production process and the provision of services and market access was disrupted. The problems were further complicated as land restitution concentrated land ownership in older households. Young and dynamic people left the rural areas in search for better opportunities in the urban areas, and often abroad, reinforcing the share of older, low skilled and less educated people in rural areas.

Pattern 4 : Large CIS countries: little restructuring and decline before 1998, recovery afterwards The land privatization processes in several of the large CIS countries were slow and because they were implemented through distribution of shares, were instrumental in keeping large farm structures in place. While this resulted in a partial maintenance of the social functions of these large-scale farms, it was an impeding factor for productivity gains. This might help explain why rural-urban differences are larger for income than for non-income poverty in Russia and Kazakhstan. Land access to household plots played a buffer role even in these countries. E.g., even in Ukraine and Uzbekistan, where no large-scale de-collectivization had occurred, household plots accounted for more than 25% of rural incomes. In Russia and Kazakhstan households produced more than half of agricultural output by the end of the 1990s. Figure 6 shows that output and yields rebounded after 1998 in the large CIS countries. Indeed, after the negative shock of the 1998 financial crisis, the associated devaluation caused an important improvement of the competitiveness and the profitability of domestic agricultural producers who saw relative prices rise substantially. Strong growth in incomes and government revenues, partly due to strongly increased incomes from mineral resources, resulted in a strong improvement in the government's fiscal situation. As a consequence, pensions, other social payments, and public sector wages have been paid and transfers to hospitals, schools, and other organizations have improved. The increase in social transfers helped to reduce poverty in both rural and urban areas.

Large-scale farms continue to dominate the land markets. Corporate farms rent large amounts of land from households, sometimes under the form of shares. More recently, corporate farms, or larger holdings, acquired vast amounts of land in ownership either through bankruptcy proceedings or, in Kazakhstan, through government policy, which transferred property rights from households to farms. These developments have constrained the income flows to poor rural households through rental markets, and have therefore limited poverty reduction.

Chapter 12

Agriculture and the New Challenges of Development

The information presented in this article was taken from the report "The State of and Outlook for Agriculture and Rural Life in the Americas, 2007: Agriculture and the new challenges of development." That document assesses the progress made in the hemisphere in achieving the four strategic objectives that the Ministers of Agriculture set in the AGRO 2003-2015 Plan, related to competitiveness, equity, sustainability and democratic governance. It also describes the main challenges facing agriculture in the hemisphere in light of the new conditions with which development efforts now have to contend.

The report concludes that, in economic and social terms, the hemisphere's agricultural sector is performing better than it has for the last 25 years, with major growth in agricultural production and trade, progress in reducing poverty, improvements in income distribution in some countries and falling unemployment rates. The outlook is positive, especially given the upturn in the international prices of core commodities, the alternative uses being made of agricultural production and promising market niches. However, there are still concerns regarding social conditions, the environmental situation and governance in rural territories.

Agriculture has played an important role in recent years, not only because it is a way of life for millions of people in the Americas but also because it is a strategic sector that contributes to the economic, social, environmental and political development of the countries.

Globalization and change. Like every other productive sector, agriculture finds itself having to operate within an increasingly globalized and changing environment. This context is determining the performance of agriculture and rural life, increasing access to knowledge and shortening the distances between markets. Furthermore, it is promoting the consolidation of a multilateral institutional system and creating opportunities and posing challenges for the economic agents involved.

Trade liberalization. As a development strategy, over the last two decades this has helped expand the markets of some productive sectors and given consumers access to a more diverse basket of goods. The outcome of the multilateral negotiations within the framework of the Doha Round of the World Trade Organization remains uncertain and the countries have turned to bilateral agreements instead. However, when negotiating bilateral agreements countries have to address the economic and social differences that exist between them as well as the social and environmental challenges posed by globalization.

From commodities to processed products. The food market is increasingly promoting the consumption of processed products instead of commodities, quality and specialized segments such as healthy, wholesome foods (low in fats and calories), denominations of origin and prepared or convenience foods. A case in point is the global organic products market, whose sales top US$30 billion, with some 1500 articles available and nine percent annual growth. This trend is a great opportunity. On the other hand, the biggest challenges are posed by the small number of actors - mainly large supermarket chains that are imposing new and greater demands in terms of volumes, deadlines, deferred payment, price setting, labeling and packaging - and the need to integrate small and medium-scale producers effectively into value chains so they can take advantage of market niches.

Technological change is providing new opportunities. The intense use of know-how is the chief characteristic of the current paradigm. The most striking development is the growing use of genetically modified organisms and their impact on certain crops such as grains and oilseeds.

Another major development is the emerging use of agriculture as an alternative source of energy, with sugar and grains being used to produce ethanol and biodiesel. In both cases, production could be increased and more income generated in rural areas, but a number of important factors must also be taken into consideration:

a. Health and the ecological balance,

b. The effects on the prices of inputs and food,

c. Food security and crop substitution, and

d. The inclusion of the production of family agriculture.

The danger of countries becoming ungovernable. The growth of international drug trafficking and its adverse effects on economic, social and political life are fueling corruption and undermining governability. The continued existence of rural poverty in most of the less developed countries, the poor quality of the jobs available and the difficulty of improving the distribution of wealth increase the risk of countries becoming ungovernable. Migration is on the rise as people seek to escape from poverty and a hand-to-mouth existence in the rural milieu. On the other hand, migrants are sending growing amounts of funds back to their places of origin. This income is helping to improve the situation of many rural households but efforts are needed to promote the use of this capital for productive investments, mitigating the problems of family disintegration and generating dignified jobs that discourage people from migrating.

Environmental degradation is increasing There is evidence of a serious deterioration in natural resources. Climate change is having a big impact on agriculture, biodiversity and human health. Greenhouse gas emissions have not been controlled and the pollution of drinking water is on the increase.

The manifestations of climate change (variations in temperature, precipitation, water levels and the increasing number of extreme events such as hurricanes, droughts and floods) will have a serious effect on agricultural competitiveness. The danger of forest fires is increasing, soil erosion is intensifying and supplies of freshwater are at risk (increasing the likelihood of more desertification). Achieving sustainable development calls not only for greater awareness of the need to manage natural resources and the environment correctly but also the adoption of good productive practices in expanded agriculture.

The conceptual framework of the AGRO Plan and the approach it adopts for the region provide the frame of reference for analyzing the recent performance of agriculture in the Americas and the impact of global trends.

The Third Summit, held in Québec was a defining moment for agriculture and rural life in the Americas. In the Declaration adopted at the Summit, the Heads of State and Government acknowledged that agriculture was a way of life for millions of rural dwellers and a strategic sector of the socioeconomic system that generates wellbeing.

In response to the mandates issued by the Heads of State and Government in the summit process, the ministers of agriculture adopted the AGRO 2003-2015 Plan, thereby placing agriculture and rural life on the inter-American agenda and providing a guiding framework for determining national and regional strategies and analyzing the sustainable development of agriculture and the rural milieu.

In preparing the 2007 report, the productiontrade, ecological-environmental, sociocultural- human and political-in-stitutional dimensions were adopted as categories of analysis in assessing the degree of progress made in achieving the strategic objectives of agriculture and rural life..

Agrifood exports are growing but now account for a smaller share of the global market. The Americas continue to be the world's leading net exporter of agricultural products, followed by Oceania. It is interesting to note that Africa, historically a net importer of agricultural products, has had a surplus since 2004. The other regions of the world continue to have a negative agricultural trade balance. The hemisphere's share of global agricultural trade has been declining since 2000. This is because its agricultural exports have grown more slowly than those of the rest of the world. While global agricultural exports grew at an annual rate of 10.5% between 2000 and 2005, this hemisphere's agricultural exports rose by only 7.1% per year.

While the share of agricultural commodities in world trade has fallen, down from 13% in 1995 to 7.2% in 2005, the share of processed products has risen 10.8% per year since 2002. The Americas mainly exports commodities, with soybean leading the way (8.5% of all agricultural exports). On the other hand, the region mainly imports processed agricultural products, a trend that has become more marked in recent years.

Agriculture in the Caribbean is becoming less competitive. Measured in terms of its revealed comparative advantage (RCA), the competitiveness of the hemisphere's agricultural trade was positive and improved during the period 2000-2005. This was mainly due to the sustained increase in the agricultural RCA of the Southern Region, since it remained unchanged in three of the other regions (Northern, Andean and Central) and fell in the Caribbean, where it actually dipped into negative territory in 2004.

Spike in international prices. During the same period, international prices of agricultural products rose sharply, by an average of 6.5% per year, reversing the trend of low prices seen in previous years. With the exception of cocoa beans, the international prices of all the main commodities rose (coffee and sugar prices alone increased by over 13%). Since the prices of tropical products increased the most, the outlook for further growth in the demand for food is positive.

Food health and safety are key elements of competitiveness. As already mentioned, quality products are now required to meet tougher standards. In 2005, nearly US$190 billion in agricultural exports and US$130 billion in imports had to meet a sanitary or phytosanitary standard of some kind.

The competitiveness of specific agricultural products could be affected by the impact of problems such as foot-and-mouth and mad cow disease, resulting in the loss of markets and jobs, falling consumption, lower prices for producers, etc. Therefore, it is important for countries in the Americas to focus on sanitary issues.

Although LAC invests over US$10 billion per year in science and technology, it is not enough, because the technological gap between LAC and the rest of the world is widening. For example, yields per hectare in LAC remain lower than those of the industrialized countries and emerging regions such as East and Southeast Asia.

Although LAC is trying to incorporate alternative production methods (such as direct planting and precision and organic agriculture), it remains a long way behind other regions of the world. Other opportunities such as agroenergy production, agrobiotechnology and agrotourism are enhancing agriculture's role as a driving force for growth in rural territories and as an agent of economic change. For example, the growing use of genetically modified varieties in this hemisphere and other parts of the world has shown that it is possible to increase yields and lower costs. It is up to the countries to ensure that producers and consumers enjoy the benefits of biotechnology, not only the developers of the technology.

The pertinent regulatory frameworks must also be created and precautions taken to prevent possible risks. With fossil fuel prices expected to continue to rise, driven by the growing demand for energy worldwide, the countries are promoting the production of fuels from agricultural products. Although many countries are endorsing these initiatives (with the United States and Brazil most committed), the impact on agricultural prices, crop substitution, food security, feed for animals and other issues remains unclear.

Furthermore, the increased demand for raw materials used in the production of biofuels is expected to alter the terms of trade for many agricultural commodities.

Poverty and extreme poverty in rural areas decreased during the first half of this decade. In 2005, for the first time in 25 years the rates for both dipped below 1980 levels. Although LAC continues to be the most inequitable region in the world in terms of income distribution, between the end of the last decade and the middle of this, the situation in rural areas improved in 11 of the 13 countries for which information is available. The gaps between poverty and extreme poverty in rural and urban areas narrowed.

Paraguay, El Salvador, Mexico and Brazil made the biggest strides in reducing poverty during that period. Uruguay, Brazil, Mexico and Panama were also on track to achieve the poverty reduction target set as part of the Millennium Goals, which only Chile has met so far. Not only are poverty rates higher in rural areas than in urban areas, the incidence of rural poverty is also higher among self-employed agricultural workers. This suggests that, as an economic activity, agriculture needs to focus on poverty alleviation strategies, because it has the greatest potential for raising the incomes and improving the quality of life of the poor.

In 2005, the remittances that migrant workers sent back to LAC had a significant effect in reducing poverty, as they totaled US$48.3 billion. Mexico receives the largest slice of these resources (US$21.8 billion per year), followed in the region by Colombia and Brazil. Remittances play a particularly important role in other Central American and Caribbean countries.

Environmental degradation is continuing and vulnerability is increasing. LAC has 915 million hectares of forests (23.2% of the world total) but 4.3 million hectares are lost every year. The United States, Chile, Uruguay, Costa Rica and Cuba are the only countries whose forest cover has increased.

Enormous biodiversity and genetic riches exist but habitat destruction is leading to the extinction of a growing number animal and plant species. Marine resources are being overexploited and desertification and the degradation and loss of soils are on the rise, due to inappropriate practices, deforestation and climate change.

The Americas contain 45% of the world's entire supply of freshwater. Despite its relative abundance, however, water is a very threatened resource (agriculture uses nearly 69% of the total). In recent years, the region has become increasingly vulnerable to natural disasters, with the poorest segments of the population and rural areas hardest hit. The Andean Region accounts for over 50% of the deaths caused by natural disasters in the Americas, with floods posing the biggest threat to human life. The Central and Caribbean regions are next in importance.

The pressure to meet the dietary needs of a population that has doubled in the last 50 years has led to nearly 37% of all land being used to grow crops, with most production systems geared to the non-sustainable use of natural ecosystems. The environmental costs have been huge in terms of the degradation of water resources, the non-sustainable use of aquifers, the loss of forest cover, desertification and the release of large quantities of greenhouse gases into the atmosphere. In general, economic development is being achieved at a burgeoning environmental cost.

Climate change is impacting the region's agriculture. The effect of climate change on agricultural production will vary from one area to another, sinçe some developed countries at mid and high latitudes, especially in the Northern Hemisphere, could actually experience net gains, while many developing tropical countries could face even bigger problems related to climate change and the increasing unpredictability of rainfall levels.

Governance in the LAC countries has been strengthened. This has occurred thanks not only to increased citizen involvement in the formulation and execution of development policies, but also because the markets are helping to achieve social objectives such as the reduction of poverty and inequity. However, the social divide threatens governance in many rural territories in LAC.

The countries are opting to negotiate bilaterally. With the Doha Round stalled and the negotiations for the Free Trade Area of the Americas suspended indefinitely, the LAC countries have opted to strengthen their respective regional trading blocs and/or negotiate bilateral or regional free trade agreements with the United States and other major economic powers (mainly the Asian economies).

And reducing the amount of resources they allocate to the countryside. Finally, the amount of public funds allocated to the rural milieu has decreased substantially. Average per capita spending on rural areas in LAC fell by over 36% in real terms over the last two decades, and total public spending in rural areas continues to be considerably less than the contribution that agriculture makes to economic development.

Making agriculture more competitive

In an environment of globalization and opening, the agricultural sectors of the LAC countries have to become more competitive. To do so, agriculture must produce more efficiently, increase the productivity of the factors of production, generate higher quality products, and have better institutions and human resources and more effective public policies geared to agrifood chains and rural territories. To produce more efficiently, with higher yields and productivity of the factors of production, efforts are needed to reorient and speed up technological change and innovation in agriculture. More and better agricultural entrepreneurs are also required, as well as a more skilled workforce and improved delivery of the services that agriculture needs to compete in more open and competitive markets. There is an urgent need to increase investment in agricultural research and development and speed up the incorporation of modern technologies into agriculture and agribusiness. The challenge facing agriculture in our countries is complicated, as we must produce more food and raw materials while at the same time conserving the forest cover and ecosystems.

To turn out quality products and compete successfully, the countries must make agricultural health and food safety (AHFS) a key element of their agrifood chains. This calls not only for improved national services but also for a new technological infrastructure and consensus with private sector organizations and actors on long-term agendas.

Formulating and implementing more effective policies in support of agriculture calls for the production of public goods and the creation of the conditions needed for them to operate properly infrastructure, information systems, legal framework, financial system, logistical system, trade treaties, etc. At the same time, efforts must be made to improve the integration of small producers and the family economy into agrifood chains and international markets.

Achieving greater equity in agriculture and the rural milieu

This calls for corrective actions to improve the traditional performance of the actors in agriculture and the rural milieu (enhancing Four challenges facing agriculture and rural life capabilities, increasing investments and promoting institution building).

Such corrective actions should form part of policies designed to promote rural prosperity that focus on four areas: agrifood chains (creation of agribusinesses, greater inclusion and job creation); rural territories (systemic competitiveness, reduction of transaction costs and generation of social income); institutional innovation (participatory and inclusive policy management); and the development of strategic capabilities (effective implementation of policies). Food security must also be made one of the key objectives of poverty alleviation strategies, technological innovation processes, international trade negotiations and the orientation of development aid.

Making the transition to a more sustainable paradigm

To meet this challenge, LAC countries must review and change the ways that soils and forests have traditionally been used for agricultural activities, and then develop competitive business strategies that attach value to, and recognize the importance of, sound environmental management and promote equity for rural actors.

The use of agriculture as an alternative source of energy could stimulate the creation of agribusinesses, jobs and income in many rural territories. It could also reduce environmental pollution, striking a balance among the need to produce food, the expansion of agroenergy crops and the pressure on forests and protected environments.

The strong growth of organic production could also make an important contribution to the transition to a sustainable development paradigm.

The countries should also take advantage of the benefits of environmental functions and services such as integrated landscape management, promoting intersectoral and participatory approaches to territorial management, facilitating economic mechanisms whereby a fair price is paid for the functions of ecosystems and fostering comprehensive territorial management.

It is also imperative to tackle the challenges posed by global climate change by adopting the following practices:

- Actions and policies that promote the generation of

accurate scientific information

- Analysis of the opportunities and risks for agriculture and the rural milieu

- Greater control of polluting emissions - Analysis and assessment of the positive and negative impacts of producing biofuels

Although all nations will be affected, the tropical countries and their poor rural inhabitants are the most vulnerable in terms of potential victims and capital losses. It is essential, therefore, that agricultural producers adapt and the national authorities prepare the rural population for the challenges posed by climate change.

Promoting governance in rural territories

To strengthen governance in rural territories, the countries need to bolster the market as a mechanism for growing the agricultural and non-agricultural rural economy. At the same time, they should enhance the capabilities of agroindustrial producers and family farmers so they can integrate better into international markets. Agriculture and the rural milieu will then be able to obtain greater benefits from trade and from the emergence of new market niches.

Furthermore, mechanisms need to be developed to improve the distribution of profits throughout agrifood chains and promote a fair return for each factor of production. The citizenry should play a bigger role in both the formulation and or implementation of public policies designed to strengthen rural territories, and the search for solutions to local problems.

Final thoughts

The role that agriculture has played in reducing poverty in recent years is a reaffirmation of the concept that IICA has been promoting since the turn of the century, namely that agriculture is important not only because it is a way of life for millions of people in the Americas, but also because it is a strategic sector that contributes to national economic, social, environmental and political development, and in some countries is still the main driving force of the economy.

Agriculture's true contribution to economic development in the hemisphere is enhanced when the region's favorable agroecological conditions are underpinned by public policies that make its products more competitive, promote trade negotiations that guarantee improved market access and make it possible to take advantage of new market niches. Also needed are more effective public policies that guarantee product quality and safety, allocate more resources to investment in science and technology and promote equity and sustainability. Actions of this kind are essential if agriculture is to be the catalyst for translating the current sustained economic growth in the developing countries into significant reductions in poverty and hunger.

The increased liberalization of agricultural trade and the expected growth in the global demand for food are a great opportunity for the region's agriculture to make an even bigger contribution to the development of the countries of LAC.

Climate change, agricultural policy and poverty reduction how much do we know?

Projections suggest that, by the end of the 21st century, climate change could have had substantial impact on agricultural production and thence on the scope for reducing poverty. This paper seeks to trace the likely impacts through changes in the quality of the physical asset base, access to assets, and impacts on grain production and on agricultural growth more generally. At moderate degrees of warming, impacts are likely to be negative in some regions, but positive in others, making it important to understand the possible implications for trade between the regions. The short term impacts of climate change, particularly changes in the frequency and severity of adverse weather events, remain uncertain, but their impacts on many developing countries are likely to be negative. There is likely to be time to make appropriate policy responses to some of the longer-term impacts.

Considerable uncertainty surrounds long-term patterns of climate change and their likely impacts on agriculture. The prospects are that policy will have time to respond to some impacts, such as possible global decreases in crop production.

Shorter term shocks from increased climate variability might be experienced much sooner and are likely to be severe for tropical areas.

Both long- and short-term changes imply an increased need for more food trade between the OECD and rest of the world. Without this, food security in some regions may diminish.

Agricultural practices need to be incorporated into mitigation policies and programmes such as the Clean Development Mechanism. This will promote the measurement of carbon balances in agriculture and the search for synergies between mitigation and adaptation in the sector.

Increased scope for carbon trading may in some measure compensate for lost agricultural production potential in developing countries if these countries can access the carbon markets. Conversely, if not carefully designed, mitigation policies such as biofuels and carbon markets are likely to further restrict access by the poor to productive resources, access which is already under threat from numerous development pressures on resources Donors can help in mainstreaming policy responses to climate change into poverty reduction strategies and into other national development policies and programmes.

The "mainstreaming" of responses to climatic change into wider government policy will help in identifying which rights of the poor are under threat and in strengthening them.

Donor assistance may usefully support developing country interests in international climate negotiations, and help governments to support both regulated and voluntary carbon markets. Specific international funds for agricultural adaptation need to be identified that are additional to existing development assistance, possibly targeted towards viable export options, and if necessary earmarked for access by developing countries.

Natural Resource Perspectives

For present purposes, climate change is defined as a process of global warming, in part attributable to the 'greenhouse gases' generated by human activity. Accompanying changes are likely to be both global, as with rising sea levels attributable to ice-melt, and local, such as changes in rainfall patterns. Responses to climate change can either seek to reduce the level or rate of change or manage its consequences. We are concerned here with both types of response.

Agriculture currently accounts for 24% of world output, and uses 40% of land area. It is highly dependent on the climate and human dependence on agricultural livelihoods, particularly by the poor, is high, and so agriculture has been a focus of those modelling the impact of climate change on poverty.

This paper reviews current knowledge about the relationships between agriculture and climate change, focusing on: cereal crops rice, wheat and maize make up 85% of world cereal exports, and are thought to be particularly sensitive to climate change four scenarios of future climate change derived from models of the Intergovernmental Panel on Climate Change and modelling studies used in the recent IPCC Working Group II Report *global* climate changes, as these have more coverage in modelling studies, rather than regional or local changes.

Most models suggest that climate change will slow or reverse the poverty reducing impact of agriculture, with, by one estimate, some 600 million additional people at risk of hunger if temperature increases by over 3°C. But this has to be set against other changes in agriculture such as improvements in technology and changes in farming systems.

Also, assumptions about population growth and food demand have a large influence on future projections.

Given the complex relationships between crops, atmospheric composition and temperature, combined with the complexities of world agricultural policies and trade, to make predictions about the future impacts of climate change on agriculture is fraught with difficulties.

Global climate models

These seek to link predictions of future climates to potential impacts on crops. There are five main elements to these processes:

Scenarios of greenhouse gas and aerosol emissions

Scenarios of greenhouse gas concentrations

Projections of temperature changes

Impacts on agriculture generated through, for example, crop response models

Impacts on agricultural trade investigated by linking these models with agricultural production, demand and trade models

Impacts of climate change on the physical asset base

Modelling of environmental constraints in relation to agriculture starts with a current situation in which two-thirds of the global land surface some 8.9 billion hectares suffers severe constraints for crop cultivation: 13.2% is too cold, 26.5% is too dry, 4.6% is too steep, 2.0% is too wet, and 19.8% has poor soils. For e.g. southern Africa by 2080, climate change could have made an additional 11% of the total land area severely constrained for crop cultivation constraints were in relation to soils, terrain, cold temperatures and moisture.

Increased carbon dioxide concentrations can have a positive impact on plant growth, although the effects are very uncertain e.g. at one extreme the projected number of people at risk of hunger in 2080 are at levels similar to 2000, at the other, almost twice that number. Increased frequency of extreme weather events could also depress yield by damaging crops at key growth stages. However, agricultural land suitable for cereal crop cultivation could expand significantly in North America (40%), northern Europe (16%), the Russian Federation (64%), and East Asia (10%), due to longer planting periods and generally more favourable growing conditions under warming.

Impacts of climate change on access to assets

The claims made by non-agricultural development on scarce land and water are set to increase. These will tend to exacerbate the insecurity of tenure which the poor already face in relation to land and water resources. This trend is likely to worsen as claims on resources shift with e.g. the increased financial flows associated with mitigation instruments CDM and emissions trading, Reduced Emissions from Deforestation and Degradation, biofuels and voluntary carbon schemes outside Kyoto's market mechanisms.

An appropriate balance needs to be struck between governments' international mandates and the pro-poor dimensions of national policies. This can be done by:

Mainstreaming climate change into development policy via Poverty Reduction Strategy Papers the United Nations Framework Convention on Climate Change National Adaptation Programmes of Action (NAPAS); and Sector Wide Approaches.

Adopting a rights-based approach in order to:

- o Strengthen the negotiating position of the poor
- o Reinforce their assets base

Case studies of land tenuremanagement and water resources indicate the significant challenges that remain to pro-poor development, challenges which derive as much from the political economy of resource exploitation as from the nature of the climatic constraints. However, one positive message is that a rights-based approach, coupled with improved understanding of likely patterns of climate change, offers a valuable opportunity to pursue longer term policies than are normally evident in conventional mechanisms such as PRSPs and SWAps. Only when the rights of the poor are more fully established and acknowledged will there be some prospect of tailoring the impacts of climate change to the wider benefit by 'mainstreaming' it within development policy. This is also likely to enhance the capacity of communities to cope through more robust 'adaptation' options, as well as addressing the wider international interests that are likely to arise with increasing international investment in climate change adaptation and mitigation.

Impacts of climate change on global cereal production

Fischer et al project that global cereal production could continue to increase up to 3.7-4.8bn tonnes by 2080 without climate change. When it is factored in, global cereal production could be within 2% of reference scenarios, but with potentially large regional variations. In general, decreases are expected in low latitudes and developing countries, reflecting both declining potential land available for crop cultivation reported above and changes in productivity. Sub-regional variations are masked by these figures, with some short term increases possible in areas of overall decrease. For example, in tropical highlands where current low temperatures prevent planting of certain crops, new land could become suitable for agriculture.

Prices and agricultural GDP

At high degrees of warming (>5°C) some models project price increases of up to 30% on average, though most projections are generally more modest, and in the short and medium term real prices could fall owing to higher outputs from slight temperature increases. Also, the impacts of climate change on agricultural GDP until 2080 are likely to be small at global level, and range between -1.5 to +2.6%, depending on the scenario, but with decreases in most developing regions.

Impacts on trade

Simulations based on the SRES scenarios outlined above project an approximate 25% increase in developing countries' net cereal imports, totalling between 170 million tonnes and 430 million tonnes in 2080, depending on the scenario. Some support within developing countries to sustainable resource management may make soil and water more resilient against climate change impacts and so help towards diversification. Public expenditure on agricultural research and advisory services, market development, and rural infrastructure can be shifted in favour of small farm export promotion. Help in diversifying out of agriculture, and more social protection, will be necessary in those areas, or for those farms, without successful farming futures.

The private sector potentially also offers new market opportunities, for example through carbon trading, the value of which could exceed official development assistance. Donors can help to build developing country capacity in international carbon negotiations as well as helping them to support both regulated and voluntary carbon markets.

Trade will impact on poverty by: altering the total supply of food within a country from the combined effects of changes to domestic supply, to the volume of exports and to the relative price of imports and exports; altering the balance of production between those currently net producers and those who are net consumers, between those producing for the domestic market and for export, and between those who source their consumption domestically or from imports; altering the structure of production between small and large farmers and between landed and landless labour.

Specific technical features of production will influence how agricultural production and trade evolve under climate change. But it is possible to identify policies both national and international that would tend to facilitate a particular outcome or the reverse.

Impacts on food security

Climate change may slow rates of improvement in food security, although the projections are highly uncertain due to limitations in the number of crop and economic models available and simplification of the definition of food *security* to food *availability*. It is projected that in 2080 around 1300 million people could be at risk of hunger under the most extreme scenarios, with the poorest countries worse affected. In these, a large portion of the population will continue to depend on agriculture and capacities to adapt to climate change, both at national and farm level are lowest. Within sub-Saharan Africa the negative impacts are likely to be strongest in north and south, possibly with some positive impacts in central African countries. General modelling studies on food security rarely consider how it could be disrupted by more extreme weather events.

Under more moderate scenarios, climate change appears to have a negligible effect on the numbers of people at risk of hunger.

Climate change, growth and poverty reduction

Against a resurgence of donor interest in the contribution of agricultural growth to poverty reduction, there are three critical areas of debate, yet concerns over climate change have been articulated into these only to a limited extent:

Are optimal combinations of growth and poverty reduction more likely to come from small rather than large farms

One emerging view is that small farms have proven to be "good adapters" by e.g. responding to weather as crops are planted and mature. But they cannot adapt in the longer-term unless they can access micro-insurance, credit, new technology and market information. In reality, larger farmers are better able to take risks, and to support small farmer adaptation through longer-term policies is likely to be difficult.

Will a focus on cash or food crops offer the best prospects for growth and poverty reduction

The evidence is scant, but, where cash crops are perennial, producers find it difficult to adapt. To reduce risk, food crop producers adapt by diversifying their range of crops, switching to drought-resistant crops and possibly even withdrawing from markets. Carbon sequestration may generate new "cash crop" opportunities – i.e. selling for income from carbon markets, as may crops for generating biofuels. This may imply a shift towards tree crops and or agroforestry, or possibly new cropping systems, such as reduced tillage.

Will investments in high or low potential areas offer the best prospects for growth and poverty reduction in agriculture

The prevalent view is that low potential areas are highly vulnerable to climate change and significant investments will be needed to maintain production there. An alternative to this type of investment is to support diversification into the rural non-farm economy and or to enable outmigration. In some zones, high potential areas which are double or triple cropped will be most under threat. Where food security or export earnings are threatened, governments may choose to protect these high potential areas rather than invest in marginal areas.

Knowledge gaps

There are large uncertainties in current projections. These result from gaps in the science of climate change, in uncertainties over crop responses, in complex socio-economic relationships and in the lack of detail in current models. For instance, in relation to rainfall prediction in Africa, there are several things we do not know which are important for agriculture, including the time of onset of seasonal rainfall and the prevalence of dry spells within seasons. Even in 15 years' time, models are unlikely to be able to project the impacts of climate change at scales of less than 50km. If this is the case, policymakers might best proceed via more 'flexible' policy approaches to cope with increased uncertainty about possible futures.

Conclusions

Shock events driven by climate chance are likely to start occurring in the near term and will require increased disaster preparedness as well as increased flexibility in e.g. trade policies. At a global scale, more gradual impacts on yields are unlikely to be felt if at all for several decades. This allows time for agriculture-focused policies to adapt.

Thus, in places where climate change will mean new markets and productive opportunities, to increase productivity now and make small farmers more efficient will place them on a strong footing to take advantage of climate change.

Elsewhere, climate change may provide a short-term window of opportunity for specific activities in the agricultural sector for example making the most of regional and domestic grain trade opportunities whilst they still exist.

Most importantly, countries that have both diversified economies and strong agricultural sectors will fare better under climate change scenarios. Thus, development assistance for the next few decades should be focused on investing in countries in the earliest stages of development. Getting an enabling environment in place, getting markets working, putting social protection in place, and strengthening R&D will enable the agriculture sector and agricultural livelihoods to be more resilient, as well as stimulating wider economic growth. There is a need for improved coordination between climate change modellers, agricultural economists and agricultural policy-makers, in order to integrate issues of climate change into existing agricultural policies and programmes. In the future, there will be a need to incorporate agricultural practices into mitigation policies and programmes such as the Clean Development Mechanism. This will lead to measurement of the carbon balances of both agricultural and wider development activities and the search for synergies between mitigation and adaptation in the agriculture sector.

Chapter 13

Organic Agriculture, Poverty Reduction, and the Millennium Development Goals

Introduction

Of the world's 1.09 billion extremely poor people, about 74 % or 810 million live in marginal areas and rely on small-scale agriculture for their livelihood. In most developing countries, agriculture continues to be the most important sector of the economy, accounting for the biggest proportion of employment. As such, unless effective strategies for agriculture development are successfully implemented, ending rural poverty will remain a distant goal.

During the past few decades, the Green Revolution has brought about significant changes in the world's food production systems. It is recognized that while the Green Revolution has benefited better-off farmers in irrigated areas, it has by-passed the poor in marginal areas. Low-external input sustainable agriculture has long been viewed as an alternative for areas where the Green Revolution technologies are not feasible. More recently, one particular alternative that has gained interest is organic agriculture, due to its commercial viability. Most farmers in marginal areas practice traditional agriculture methods using very little or no agrochemicals. By adopting organic agriculture (OA), which requires less financial inputs while placing more reliance on natural and human resources, farmers could move towards more sustainable agricultural practices. Improving the agricultural production system in marginal areas in a sustainable manner and providing market access for the poor hold the key to the mass reduction of poverty.

Recognizing that the needs of the world's poorest are multidimensional and recognizing the needs for people-centered, time-bound, measurable indicators to monitor effectiveness of poverty reduction efforts, the states of the United Nations have agreed to adopted the eight Millennium Development Goals to assess poverty reduction efforts. The MDGs is a set of eight goals derived from the various declarations and commitments adopted in conferences organized by the United Nations. The MDGs cover a diverse set of development outcomes, ranging from halving extreme poverty to protecting the environment and promoting a global partnership in development. Each goal has corresponding time-bound targets, most of which should be achieved by 2015. The MDGs have been universally accepted as a framework for measuring development progress, and have become a blueprint for development interventions adopted by UN member countries and all the world's leading development institutions. At both the national and international levels, policymakers and donors are now evaluating development interventions and strategies in relation to their impacts on the MDGs, with a goal of achieving the targets by 2015.

Agriculture's potential contribution to the MDGs has been widely recognized. More recently, international organizations such as the UN's Food and Agriculture Organization the International Fund for Agricultural Development, the International Food Policy Research Institute and the World Bank, have produced a number of discussion papers on agriculture and the MDGs. In this paper, we look specifically at how organic agriculture can contribute to the MDGs. 1 The International Federation of Organic Agriculture Movements defines organic agriculture as a "holistic production management system which promotes and enhances agroecosystem health, including biodiversity, biological cycles, and soil biological activity. It emphasizes the use of management practices in preference to the use of off-farm inputs, taking into account that regional conditions require locally adapted systems."

This paper reviews the existing literature on organic agriculture, poverty reduction and its linkages to the Millennium Development Goals. The paper aims at informing governments and donors of the potential impact of OA on the MDGs along with identifying the key gaps in the knowledge base on the linkages for each MDG. Ultimately, the paper hopes to provide a starting point for developing a research agenda that can be used for policy formulation to support organic agriculture, particularly in marginal areas where majority of the poor reside.

This paper is organized into five parts. Following this introduction, Part II looks at the status of rural poverty in Asia and explores the factors behind the increasing interest in OA as an alternative strategy for poverty alleviation in marginal areas. Part III reviews the existing literature on the impact of OA and attempts to link these impacts to the MDGs. Based on this literature review, Part IV concludes.

In cooperation with developing countries, develop and implement strategies for decent and productive work for youth.

Organic Agriculture as a Strategy for Addressing Rural Poverty in Asia

A. Rural Poverty In Asia

In 2003, close to 700 million Asians were estimated to be living on less than a dollar a day. Poverty in Asia is largely a rural phenomenon: although the magnitudes and ratios would vary across countries, almost everywhere in the region income poverty is disproportionately higher in rural areas.

Rural areas are generally worse off in almost every other aspect of poverty and deprivation: people in rural areas tend to have lower levels of health and education; they are more likely to have limited access to basic services such as water and sanitation; and paradoxically, despite depending on agriculture as their main source of livelihood, they also suffer the most from hunger and food insecurity. The rural poor are also the most affected by environmental degradation: fifty percent of the poor in Asia are found in fragile ecosystems and vulnerable areas, and their livelihoods largely depend on the available natural resources.

Environmental degradation reduces the income of the rural poor disproportionately in comparison with the general population. Perversely, as the incomes of the poor shrink, they are left with little recourse but to deplete environmental resources even further, thus precipitating a vicious cycle of ever-worsening poverty and environmental deterioration. This poverty-environment nexus has become prominent in the discussions on sustainable development.

While many countries in Asia may be poised to meet the MDGs in 2015, these aggregate gains are not likely to translate into better living conditions for the rural poor unless something is done to re-focus attention on rural poverty and agriculture3.

B. Green Revolution and Poverty Reduction

In the past few decades, massive investment has gone into promoting Green Revolution technologies based on the use of chemicals, extensive irrigation, and the use of high yielding varieties, including genetically modified plant varieties.

While there is no doubt that this strategy has led to substantial productivity gains over the past 50 years and has eliminated starvation in many countries, recent evidence shows that the Green Revolution has not been effective as a strategy for poverty reduction for majority of the world's rural poor. While Green Revolution technologies will remain as the major production system in the world, there is growing evidence that the Green Revolution has, at its worst, increased inequality, worsened absolute poverty, and resulted in environmental degradation.

First of all, while Green Revolution methods have been effective in increasing yields in agriculturally optimal areas, they have been less effective in the case of marginalized and resource-poor areas where farmers have no access to modern inputs and technologies.

"Even in Green Revolution regions, numerous small, poorly equipped and very low-income farms were unable to gain access to the new means of production. Unable to invest and progress, they saw their incomes fall as a result of the drop in real agricultural prices. Many of them sank to levels of extreme poverty and were eliminated. Above all, vast hilly and barely accessible regions of rainfed or scarcely irrigated agriculture were essentially bypassed by the Green Revolution. The varieties cultivated in these regions millet, sorghum, taro, sweet potato, yam, plantain, cassava benefited marginally, if at all, from selection. The same was true for varieties of major cereals wheat, maize, rice that were adapted to difficult local conditions altitude, drought, salinization, aridity, waterlogging. For example, the average output of millet throughout the world today is barely 800 kg/ha, and that of sorghum is less than 1 500 kg/ha. These so-called "orphan" varieties, having been bypassed by the selection process, make the use of fertilizer and phytosanitary inputs unprofitable, which only adds to the problems of the regions where they are grown."

Second, even in those areas which enjoyed substantial productivity gains, in the long-run these gains did not always translate into sustainable improvements in rural poverty. On the one hand, real declines in the prices of agricultural commodities and increasing crop failures due to pests and diseases resulted in a significant fall in farmers' revenues. Because high yielding varieties often need regular or increasing inputs of chemical fertilizers and pest control, farmers had to borrow heavily in order to sustain productivity.

In the long-run, this cost-price squeeze and the declining price trend of commodities in the world market led to significant declines in terms of trade and incomes of farmers.

Third and more importantly, the Green Revolution's gains have come at the cost of extensive environmental degradation and considerable health problems due to exposure to agro-chemicals. As IFPRI writes: "Excessive and inappropriate use of fertilizers and pesticides has polluted waterways, poisoned agricultural workers, and killed beneficial insects and other wildlife. Irrigation practices have led to salt build-up and eventual abandonment of some of the best agriculture lands. Heavy dependence on a few major cereal varieties has led to the loss of biodiversity on farm". And since these costs are not internalized in the price of food, it is the taxpayers and future generations who will end up footing the bill.

Because majority of the poor are illiterate and have no access to adequate training on agrochemical use, they are disproportionately affected by negative health and environmental consequences of the Green Revolution. The inappropriate use of agrochemicals and the premature introduction of mechanization have led to, in many areas, a deeper level of poverty.

With regards to health, one of the key pillars of poverty reduction, it could also be argued that the poor are in greater need of quality food to maintain or improve their health compared to the average citizen. For the poor, OA systems, which are generally more diversified, could translate into higher levels of nutrition intake. Given the above-mentioned reasons, low-external inputs sustainable agriculture strategies have emerged as viable alternatives to the Green Revolution, particularly for the rural poor in marginal areas. For farmers living in these areas, any strategy to improve agricultural production must therefore be based on the use of low-cost and locally available technologies and inputs, in addition to being safe for humans and the environment.

C. Organic Agriculture as a Poverty Reduction Strategy

The worldwide promotion of organic agriculture for poverty reduction was pioneered by farmers themselves and was advocated by NGOs, who worked closely with poor farmers and witnessed the serious negative health and environmental consequences of agrochemicals. More recently, governments and donors have taken note of organic agriculture's potential as a development strategy, due mainly to the following:

(1) increasing global demand for safe food and potential price premiums for organic products;

(2) under the WTO agreement, food exports must comply with higher phytosanitary standards and OA is more likely to be able to meet such requirements;

(3) mounting evidence that OA can improve the incomes and living standards of poor farmers by building on assets which poor farmers have, i.e., land free from intensive use of chemicals, excess labor, and traditional knowledge of production system; and

(4) studies which have illustrated how OA can contribute to health, social development and environmental restoration and or protection.

Demand for safe and organic food is rapidly expanding in both the domestic and international markets. In the urban areas of most developing countries, supermarket chains are responding to increased demand for safe food by contracting farmers to produce crops and livestock organically or with reduced amounts of chemicals.

Many of the poor farmers in remote areas possess a comparative advantage over farmers in intensive areas because the former's current practices are largely organic by default. Unlike their conventional counterparts, they will not require a transition period of 3-5 years before they can be certified organic. This gives them an edge in terms of immediately capturing the benefits of producing certified organic products for the domestic and international markets.

Organic agriculture is likely to benefit the poor living in marginal areas the most, by improving productivity and incomes, and promoting environmental sustainability. In Thailand, contract organic rice farming in marginal areas has produced significant livelihood improvements for participating farmers. In Kandy, Sri Lanka, an organic tea project for resource poor farmers has led to favourable environmental outcomes in an area where tea had almost been abandoned. Even in Europe, organic agriculture is reportedly more likely to be found in disadvantaged areas, or areas unfavorable to conventional agriculture production.

Due to declining commodity prices, countries are choosing instead to specialize in high-value crops, including organic products. Regarded as both an export opportunity and a sustainable development strategy, many Asian countries have formulated or are now in the process of formulating national policies for OA. Thailand, for example, announced the first policy on organic agriculture in 2001 and declared it as the National Agenda in 2005. Bhutan formally declared its National Strategy on Organic Agriculture in 2006. Some countries, such as Cambodia and the Philippines the promotion of organic agriculture is led by Ministry of Trade.

To the extent that international trade of socially and environmentally beneficial products such as organic products can expand, consumers in developed countries can directly contribute to poverty reduction in developing countries. Given the multi-dimensional nature of poverty and the broad-base benefits of organic agriculture to the rural poor, the following section reviews the current knowledge on OA's contributions within the context of the MDGs.

III. Organic Agriculture and the MDGs: Impacts Reported in Existing Literature

A. Goal 1: Eradicate Extreme Poverty and Hunger

Mounting concerns regarding food safety and increased awareness of negative environmental consequences of modern agriculture have led to a growing demand for organic products, particularly in developed countries. As a result, the global market for organic products has been growing steadily in Europe, North America, and some Asian countries such as Japan. Between 1998 and 2002, the compound annual growth rate of the organic food market was 17.7 percent. In 2004, the market for organic products was valued at US$27.8 billion, the largest share of organic products being marketed in Europe and North America, followed by Brazil and Middle East. The organic sector is expected to continue to be the fastest growing sector in agriculture. Much of the increase in organic production is occurring in developing countries where farmers are being attracted by export benefits and substantial price premiums. Organic products enjoy price premiums of between 10-300 percent, depending on the product and different studies estimate that farmers receive between 44 – 50 percent of the price premium.

Organic and Production Costs

Employing organic farming methods will lead to higher profits for farmers not only because of price premiums, but also because of lower production costs. OA technologies can decrease the costs of production as chemical inputs are substituted by locally available and cheaper organic inputs and more intensive labor which the poor often have in abundance. Adoption of OA systems also lowers the need for credit, which is often expensive and difficult to obtain for small farmers.

Some studies have shown that even in the absence of price premiums, farmers have turned to OA because of lower production costs. For instance, Scialabba and Hattam (2002) cite the case of rice farmers in the Philippines who adopted organic practices which allowed them to save on production costs because they did not have to purchase external inputs, at the same time achieving stable yields.

Organic and Production Risks

Studies also show that OA can lessen the risk of income losses associated with seasonal variations or crop failures. On the one hand, diversification, which is common in organic systems, has been shown to increase farm production by 20 to 60 percent as compared to a traditional low-input system. This diversity, in conjunction with greater on and off-farm biodiversity, allows farmers to derive extra income from the sale of additional products or wild crops and non-timber forest products. Some organic systems also favor the use of traditional varieties which are typically more resistant to local pests and diseases. And since OA allows farmers to save their own seeds, farmers can gradually increase crop resistance to pests and diseases by breeding these seeds for "horizontal resistance".

Anecdotal evidence also suggests that organic systems are more resistant against droughts and typhoons mainly because organic matter increases the soil's ability to take in water during rainfall events. In the Rodale Institute Farming 5 Scialabba, et.al. (undated) define horizontal resistance as" the ability of a crop to resist many or all strains of a particular pest which differs from breeding for "vertical resistance" to have a gene to resist one specific strain of a disease."

Systems Trial, OA was found to significantly improve soil structure and water capture, allowing organic farms to perform better and out-yield conventional farms by significant margins under drought conditions. The same trial also revealed that OA can reduce the potential for crop failure due to erosion in severe storms.

Organic and Profitability

There is substantial evidence linking OA with improvements in the profitability and income of poor farmers in developing countries. Case studies by UNESCAP show that certain organic farmers' groups were able to double their income due to the lower cost of organic inputs and lower credit costs. More importantly, studies show that OA has been particularly profitable for smallholders. IFAD's study in Latin America and the Caribbean reveals that OA has benefited smallholders the most. In five out of the six countries covered by the study, small farmers dominated organic production for export as well as domestic markets. In Mexico in 2000, smallholders represented 98.6% of all organic producers, accounting for 84.2% of the area under organic production and generating 68.8% of the exports of organic products.

A more recent study by Setboonsarng, et al., on organic rice contract farming likewise finds that small organic farms are more profitable and efficient than larger farms, and that in general, organic rice contract farming is more profitable than conventional non-contract farming by a significant margin for all scales of operation, even when the non-cash costs are included. Similar promising trends are reported in numerous studies by other researchers.

Organic and the Broader Economy

Organic agriculture may generate secondary effects in the broader rural economy. Lohr (2002) reports that U.S. counties with organic farms have stronger farm economies and contribute more to local economies through total sales, net revenue, farm value, taxes paid, hired labor, purchases of inputs, and repair and maintenance services. Horrigan, et al., likewise argue that profits generated by small-scale producers are more likely to remain in the community and create multiplier effects in the local economy.

Halve, between 1990 and 2015, the proportion of people who suffer from hunger

The World Food Summit in 1996 defined food security as "access to sufficient, safe and nutritious food to meet their dietary needs and food preferences for an active and healthy life." Food security is therefore a broad objective that is concerned with both the quantity and quality of people's diets.

The poor typically spend at least 80 % of their income on

food, most of it basic food staples; the poor are also at risk of consuming too few calories for health and efficiency. Although the Green Revolution may have solved food security at the national level, hunger and malnutrition still persist, particularly in rural areas. In some areas, hunger increased as cropping patterns changed from multiple to monocropping of cash crops for sale, consequently increasing hunger as cash crop prices have been declining.

In rural areas, food shortage is usually the direct outcome of crop failures due to calamities and pests. Farmers who practice monoculture tend to be the most affected when their crops fail, suffering losses in income and difficulty in purchasing food. Such problems would be less daunting in the case of farmers practicing integrated system advocated in organic farming. The organic practices of crop-rotation, intercropping and polyculture increase the availability of food throughout the year while lowering risk. The organic system often includes livestock and vegetables alongside the main crop, providing the community with improved dietary diversity and quality, two important aspects of food security.

Organic and Food Security at the Household Level

Findings from literature suggest that OA can lead to improved food security at the household level. Pretty cites the results of a study by the University of Sussex which revealed that OA resulted in improvements in food production by smallholder farmers through one or more of the following mechanisms:

1. Intensification of a single component of the farm system such as home-garden intensification with vegetables and trees;
2. Addition of a new productive element to a farm system such as fish in paddy rice - that boosted the farm's total food production, income, or both but did not necessarily affect cereal productivity;
3. Better use of natural capital to increase total farm

production, especially water (by water harvesting and irrigation scheduling) and land (by reclamation of degraded land), enabling growth of additional new dry land crops, increased supply of water for irrigated crops, or both; and

4. Improvements in per-hectare yields of staples through introduction of new regenerative elements into farm systems for example, integrated pest management or locally appropriate crop varieties and animal breeds."

These findings are supported by the experiences of OA projects in Asia, Africa, as well as in Latin America which show that diversification improves access to food as well as dietary quality.

Organic and Food Security at the National Level

The extent to which OA can contribute to food security beyond the household or community level is still being debated in the literature. Central to this debate are the issues of yield and productivity.

The impact of OA on productivity depends largely on the previous agricultural system; as noted previously, OA is reported to have the biggest potential to increase yields in marginal lands or lands under traditional agriculture. Cases from UNESCAP show that OA was able to increase the yield of those converting from traditional to OA by 200-300 percent, similar to that obtained from conventional farming.

In intensive areas, conversion to OA usually decreases yields in the first year, but by the third year, yields typically stabilize and can be almost identical to conventional yields.

However, the findings of research reviewed by Vasiliokos show that conversion to OA can lead to higher yields than conventional farming. The Farming Systems Trial at Rodale Institute likewise provides supporting evidence revealing that there was no difference in overall yields of corn, soy bean, and other crops In addition, the ISP cites data on yield changes in 89 projects which revealed that "farmers have achieved substantial increases in food production per hectare, about 50-100% for rain-fed crops, though considerably greater in a few cases, and 5- 10% for irrigated crops though generally starting from a higher absolute yield base. These projects included both certified and non-certified organic systems, and integrated as well as near-organic systems. In all cases where reliable data were available, there were increases in per hectare productivity for food crops and maintenance of existing yields for fiber." Most importantly, the yield in OA is expected to be sustainable over time.

B. Goal 2: Achieve Universal Primary Education

Rising incomes from OA may enable rural parents to increase spending on education. However, since OA is often associated with higher reliance on family labor, it could possibly have a perverse impact on education. This remains an empirical question to be assessed. In theory, the decision will depend on how the rural poor would value the opportunity cost of sending children to school rather than putting them to work on the farm. One consideration that could possibly tilt the decision in favor of keeping the children in school is the fact that modern OA, particularly certified organic is knowledge intensive and requires higher managerial and marketing skills than conventional agriculture. This could change the parents' assessment of opportunity costs of keeping a child in school. Moreover, as OA begins to have its impact on the broader economy, the demand for skilled labor in food-related sectors such as processing can be expected to rise.

C. Goal 3: Promote Gender Equality and Empower Women

In many developing countries, women play a critical role in several aspects of agricultural production and processing. In general, over two-thirds of rural women live in low-income households, and female-headed households tend to be the poorest among these, accounting for 35-40 % of all heads of households in some areas in Asia. While any improvement in women empowerment would be desirable in itself, it would also have implications on the achievement of other MDGs. Many of the MDGs will rely on empowerment of women, since women are the principal caregivers in the household. It has been argued that since AO is labor intensive, it could empower women by providing them with more earning opportunities. IFAD's study on OA in People's Republic of China shows that women in Jianxi province expressed that the ability to earn in OA provided them with a greater feeling of worth for their contribution. In many OA certification bodies in Asia, women are preferred as farm inspectors, thus expanding employment opportunities for women.

However, as Engel-Di Mauro and Dolan and Sorby emphasize, OA's impact on gender empowerment and equality will depend on initial gender relations, as reflected in gender-based divisions of labor, decision-making, housework, and intra-household allocation of resources and assets. One issue that has been extensively discussed in literature is that of time poverty: while OA may increase women's incomes, this increase may not be enough to compensate for the multiple burdens which women have to carry. It is also possible that in an effort to cope, women will shift some of the domestic burden on to other members of the household who are already vulnerable to begin with, such as the children (particularly daughters) or even other female household members who are already overburdened with work.

Another issue that needs to be investigated closely is whether women have sufficient bargaining power or role in decision-making within the household. In fact, the presence of sufficient bargaining power may have a bearing on whether OA is adopted by the household at all. The extent to which women can influence how the extra income from OA is used is also crucial as women are known to be more likely than men to spend extra income on the children and the well-being of the family.

One final aspect that needs to be taken into consideration is whether OA would lead to a reduction in, or conversely bring on, violence against women. Anecdotal evidence suggests that since converting to OA entails adopting a new farming attitude of respecting nature as opposed to exploiting nature, family harmony has reportedly increased. Experience in microfinance has revealed cases where violence against women declined as women's incomes, assets, and share in household decision-making improved. It would be helpful to investigate whether OA would have a similar positive impact. Unless all of the foregoing factors are taken into consideration, it will be difficult to meaningfully assess whether OA would have a positive impact on women empowerment.

D. Goals 4 and 5: Reduce Child Mortality and Improve Maternal Health

OA can have a direct impact on child and maternal health through lowering the health risks associated with exposure to agrochemicals, in particular pesticides. This is particularly important for farm workers in developing countries where long-term exposure to pesticides has led to serious illness such as cancer and other reproductive problems. Health damage 6 In Japan for instance, women were responsible for starting the organic movement. In the 1980s, with rapid development of industrial and service sector, men began working in the industrial sector and left the farm to be managed by women. During the same period of time, the use of green house for cultivating crops was becoming widespread but as the negative consequences agro-chemical use, particularly pesticides, became more apparent in green house agriculture, the women farmers decided to form groups to convert to organic agriculture. In the 1990s, marketing groups, generally formed by women, established market linkages with housewife associations and consumer associations.

Exposure to other agrochemicals, for example, nitrogen leaching from chemical fertilizers, has also been implicated in child deaths due to nitrate poisoning in drinking water, as nitrate hinders the proper transportation of oxygen in the red blood cells. This is know as the "blue baby syndrome". While adults are not typically affected by nitrates or nitrites, ingestion of these chemicals by babies can lead to death. In an ADB rural water supply project in Sri Lanka, communities downstream of a watershed area faced serious problems of nitrogen contamination in groundwater leading to illness and even death of babies. By promoting OA in watershed areas, the downstream villagers re-gained their access to clean drinking water while farmers upstream also improved their incomes.

Because poor farmers are largely illiterate or lack accurate knowledge about the harmful effects pesticides and fertilizers, inadequate storage and unsafe handling practices of these chemicals have been shown to widen the risk of exposure to include other members of the household, such as children .

Ransom likewise notes that studies have shown a link between a variety of reproductive health impacts in women and pesticide exposure. Increased incidence of miscarriages, birth malfunctions, still births and delayed pregnancy have been documented among women agricultural workers and wives of men employed in pesticide mixing and spraying. There is also evidence of increased risk of birth defects from parental exposure to pesticides, although the extent of this risk is uncertain.

Surveys conducted by IFAD show that health problems due to pesticide exposure is one of the main reasons why farmers who used to practice conventional farming chose to shift to organic farming. Farmers in Karnataka, India reported that symptoms associated with pesticide exposure disappeared after they shifted from conventional farming to organic farming. The IFAD study in Latin America reports that organic farming is associated with better health of farmers. The study cautioned, however, that this conclusion is based on qualitative evidence. Nevertheless, the finding shows that farmers generally perceived themselves to be healthier after converting to OA.

Looking beyond the direct impact of pesticides on farmers' health, consumers may likewise experience deterioration in health as a result of ingesting pesticide residues in food products. Between 1999 and 2003, the Ministry of Public Health of Thailand tested 4,000 food samples from various sources. Approximately half of all samples contained pesticide residues, including 45 percent of Thai vegetables and 50 percent of Thai fruit. The Ministry also found that 55 percent of imported fruits contained pesticides. Grapes and tangerines were heavily contaminated by pesticides, although residues were not detected in bananas, mangoes, jackfruits and pineapple.

Apart from lowering risk of exposures to agrochemicals, existing studies view the linkage between agriculture and improvements in health outcomes of farm households as being largely indirect: the impact is traced through an increase in income and resulting improvements in food security. Rising incomes could allow households to spend more on food, medicine, and health services leading to lower mortality rates. And as noted earlier in this paper, improvements in food security would go a long way in preventing child and maternal deaths.

Organic and Dietary Quality

Worthington a nutritionist, reports that organic crops are better at improving dietary quality:

1. Organic crops contained significantly more vitamin C, iron, magnesium, and phosphates and significantly less nitrates than conventional crops;
2. There may be less toxic heavy metals in organic crops than in conventional crops; and
3. Protein content may be less but the quality may be better in organic crops than in conventional crops."

More recently, Flock et al. cite the results of a study by the University of Washington which analyzed pesticide breakdown products metabolites in preschool aged children and revealed that concentrations of pesticide metabolites were six times less in children eating organic fruits and vegetables compared to levels found in children eating conventional produce.

Combat HIV/AIDS, Malaria and Other Diseases

OA's contribution to combating HIV-AIDS may not be clear but its contribution to combating other major diseases may be traced indirectly to its impact on the ecology. Ecological factors play a role in the spread of disease in two ways: they could allow disease vectors to become abundant, or they could suppress them. As Chivian writes, ecological disturbances "reduce the abundance of some organisms, cause population growth in others, modify the interactions among organisms, and alter the interactions between organisms and their physical and chemical environments. These disturbances have consequences for human infectious diseases whenever they influence, either directly or indirectly, the organisms involved in the maintenance or transmission of infections." As will be discussed shortly, OA helps keep these ecological disturbances to a minimum.

One way in which OA could make a contribution is through better waste and water management. Improper waste disposal such as animal matter or human waste is a major source of disease in rural areas. Since OA uses these wastes as inputs to the production system through composting, it can lead to a reduction in the prevalence of common diseases. In terms of water management, OA can reduce the risk of waterborne diseases such as malaria. For instance, irrigation projects have been implicated in large and widespread increases of schistosomiasis, and in the 1990s poorly maintained irrigation systems and uncontrolled local irrigation gave rise to "irrigation malaria," which affected about 200 million people in rural India. In fact, in some developed countries, government waterworks have encouraged conversion to OA to reduce the cost of purifying drinking water.

Bonner cites a study in Korea which revealed that avoiding pesticides in paddy fields encourages the growth of a certain fish, the muddy loach, *Misgurnus mizolepsis*, which effectively controls mosquitoes that spread malaria and Japanese encephalitis. There are also some studies on rice-fish integrated farming which provide evidence that the introduction of fish into the farming system leads to a decrease in the incidence of malaria, since the mosquito larvae are consumed by the fish. Similar experiences are also cited in the case of integrating ducks into the rice farming system.

Ensure Environmental Sustainability

Conventional agriculture has played a substantial role in environmental degradation. Stoll reports that inappropriate agricultural practices has contributed to the loss of 38 % of total agricultural land in Asia and the Pacific, while the ADB reports that agriculture has been responsible for land degradation amounting to 212 million hectares. Large-scale irrigation projects developed for conventional agriculture have consumed water at unsustainable and inefficient rates, delivering only 40 % of water to crops causing water-logging and salination. Monoculture and dependence on a few plant varieties has diminished genetic diversity: Scialabba reports that currently, 1,350 breeds face extinction, with two breeds being lost each week. Furthermore, agricultural activities affect 70 % of all threatened bird species and 49 % of all plant species.

Meanwhile, intensive agriculture's dependence on agrochemicals and non-renewable fossil fuels has accounted for over 20 % of global anthropogenic greenhouse gas emissions. In Asia, this figure could be slightly higher as most of the fertilizers used in Asia are nitrogen-based and the industrial process of producing nitrogen fertilizer releases nitrogen dioxide, a strong greenhouse gas, into the atmosphere. All of these damages have come at a considerable cost to society and the economy at large. Stoll cites figures which places the economic costs of environmental damage at between 1-9 % of GDP, depending on the country and the impacts below summarizes some of the damages associated with intensive agriculture. 10 Nitrous dioxide is a greenhouse gas that is over 300 times stronger than carbon dioxide.

Bibliography

Ahluwalia, M. S., 1978. "Rural Poverty and Agricultural Performance in India"; *Journal of Development Studies*, vol. 14 (2).

Ahmed, R. and N. Rustogi, 1987. "Marketing and Price Incentives in African and Asian Countries: A Comparision" in Dietez Elz, eds., *Agriculture Marketing Strategy and Pricing Policy*, Washington, D.C. : World Bank.

Bardhan, P., 1996. "Research on Poverty and Development Twenty Years after Redistribution with Growth", *Proceedings of the Annual World Bank Conference on Development Economics*, 1995, Washington, D.C.: World Bank. *District Level Study*, New Delhi: Sterling Publishers.

Bhalla, Surjit, 1988. "Sri Lankan Achievements: Facts and Fancy" in T. N. Srinivasan and Pranab Bardhan, eds., *Rural Poverty in South Asia*, New York: Colombia University Press.

Chadha, G. K., 1994. *Policy Perspective and Indian Economic Development: Essays in Honour of Prof. G. S. Bhalla*, New Delhi: Har Anand Publications.

De Corla-Souza Patrick, J. Everett, B. Gardner and M. Culp, 1997. "Total Cost Analysis: An Alternative to Benefit Cost Analysis in Evaluating Transportation Alternatives", *Transportation*, vol. 24, pp. 107-123.

Frankel, Francine R., 1971. *India's Green Revolution – Economic Gains and Political Costs*, Princeton.

Gannon, Colen A. and Zni Liu, 1997. *Poverty and Transport*, Discussion Paper (TWU-30), Washington, D.C.: World Bank.

Hirschman, Albert O., 1959. *The Strategy of Economic Development*, New Haven, Conn: Yale University Press.

Hossain, M. and O. H. Chowdhery, 1984. *Socio-Economic Impact of Roads in Rural Areas*, Dhaka: Centre for Development Science.

James, Douglas L. and Robert R. Lee, 1971. *Economics of Water Resources Planning*, New York: McGraw Hill Inc.

Kumar, P. and Mark Rosegrant, 1994. "Productivity and Sources of Growth for Rice in India", *Economic and Political Weekly*, vol. 29, No. 52.

Mellor, John W., 1976. *The New Economics of Growth: A Strategy for India and the Developing World*, Ithaca, New York: Cornell University Press.

Sambrani, Shreekant, 1982. "Subsidy as an Instrument for Increasing Agricultural Production and Income", Rapporteur's report, *Indian Journal of Agricultural Economics*, Oct-Dec. pp. 282-283.

Selowsky, Mercelo, 1979. *Who Benefits from Govenment Expenditure*, A Case Study of Colombia, New Delhi: Oxford University Press.

Vaidyanathan, A., 1991. "Critical Issues Facing Indian Irrigation" in Ruth Meinzen- Dick and Mark Svendsen, ed., *Future Directions of Indian Irrigation*, Washington, D.C.: IFPRI.

World Bank, 1976. *The Economic Analysis of Rural Road Projects*, World Bank Staff Working Paper No. 241, Washington, D.C.: World Bank.

Index